Agriculture and Water Management

annually [illegible] Drought or near-drought conditions also occur in parts of [illegible], Papua New Guinea, and [illegible] in India, Laos, the Philippines, and Vietnam [illegible] more frequent [illegible] following [illegible]. At least half of the severe [illegible] in India [illegible] since 1871 have occurred during El Niño years [illegible] 1998).

[illegible] common in this region. For example, in [illegible] flood-prone area is [illegible] Mha [illegible] (Jackson, 1996) [illegible] anomalous [illegible] the Eastern Equatorial Pacific [illegible] as that [illegible] the 1998 El Niño, a higher [illegible] of intense extreme [illegible] across Asia is possible.

There is [illegible] agreement among global climate models (GCMs) of a significant acceleration of warming in [illegible] in Asia over that observed in the 20th century. [illegible] that [illegible] will warm more than [illegible] approximately [illegible] IPCC TAR [illegible]

[illegible]

[illegible] precipitation [illegible] leading to [illegible]

Agriculture and Water Management

Manish Dubey

CYBER TECH PUBLICATIONS
4264/3, Ansari Road, Daryaganj, New Delhi-110002 (India)
Ph.: 011-23244078. Fax: 011-23280028
E-Mail: cyberpublicationsdelhi@yahoo.com

Agriculture and Water Management

Manish Dubey

First Edition 2009

Published by :

G.S. Rawat for **Cyber Tech Publications**
4264/3, Ansari Road, Daryaganj, New Delhi-110002 (India)
Ph.: 011-23244078. Fax: 011-23280028
E-Mail: cyberpublicationsdelhi@yahoo.com

Printed at : **Amit Enterprises**, Maujpur, Delhi, (9810878184)

Preface

Shortage of water available for agriculture is a problem in most parts of the world. Shortage and degradation of agricultural land also greatly limit agricultural development in many regions. There is additionally increasing concern that agricultural development contributes significantly to general environmental degradation. These problems have reached dangerous levels in many countries such that economic development and social structures are threatened. If future food production is to keep pace with developing population, and natural environments are to be protected, then agricultural practice must change. An increase in the efficiency with which water is used in agriculture is a major priority. For many countries, agriculture is the largest water user and to improve water use efficiency in agriculture would have a significant impact on sustainable development. It is not only important to ensure availability of water in the right quantity at the right time, but also important to ensure that water of an appropriate quality is used. Research on water efficient agriculture, and on sustainable use of water and land resources in arid and semi-arid areas has a high international priority.

The objectives of the conference were to bring together a multi-disciplinary group of researchers, engineers and regulators to present and discuss current research. The conference themes included biological mechanisms of water-saving agriculture, agronomic technology and plant improvement for water-saving in water-limited areas and dryland, irrigation technology and water management, sustainable use of water resources in arid and semi-arid areas.

Undersigned is confident enough that this work of him would be beneficial to the readers. However every possible care has been taken in preparation and printing of this book but still some errors are likely to be remained, your valuable suggestions and comments are heartily welcome.

—*Author*

CONTENTS

1

IMPACTS, VULNERABILITY AND ADAPTATION TO CLIMATE CHANGE IN ASIA

This paper was commissioned by the secretariat of the United Nations Framework Convention on Climate Change and prepared by Mozaharul Alam, Atiq Rahman, Mariam Rashid, Golam Rabbani, Preety Bhandary, Suruchi Bhadwal, Murali Lal, and Moekti Handajani Soejachmoen. In some parts of the document the secretariat introduced modifications.

INTRODUCTION

Purpose and Scope

The Conference of the Parties (COP), by its decision 1/CP.10, requested the secretariat to organize three regional workshops, reflecting regional priorities, and one expert meeting for small island developing States, in order to facilitate information exchange and integrated assessments to assist in identifying specific adaptation needs and concerns.

This paper represents background information as input to the Asian Regional Workshop on Adaptation (Beijing, China, 11-13 April 2007) and is prepared by secretariat of the United Nations Framework Convention on Climate Change (UNFCCC).

The purpose of the background paper is to provide the context of climate change vulnerability and adaptation in Asia aiming to facilitate discussion during the Workshop. The paper provides an overview of general social and environmental conditions in this region. It outlines the vulnerability of the Asian region to climate change, the current status of adaptation assessments, and the implementation of adaptation options, taking into account subregional and national differences. Asia is a challenging region because of its geographical expanse and diversity, political circumstances and varying levels of implementation of adaptation action.

The paper is intended to provide a starting point for discussions on the various aspects of impacts, vulnerability and adaptation in the Asia Region.

Sources of Information

This background paper is prepared based on available secondary data and information, including:

(a) National Communications (NCs) to the UNFCCC, National Adaptation Programmes of Action (NAPAs), materials from previous UNFCCC regional workshops e.g. those undertaken by the Consultative Group of Experts on National Communications from Parties not included to the Convention (CGE) and the Least Developed Countries Expert Group (LEG), and other relevant UNFCCC documents.

(b) The Intergovernmental Panel on Climate Change Third Assessment Report (IPCC-TAR), as well as more recent publications on climate change (especially from peer-reviewed journals), literature on sustainable development and other relevant topics including outcomes of relevant conferences held in and for the Asia region.

(c) Where available, government strategy papers for the consideration of climate change in planning and development, for example in the improvement and expansion of sectors such as infrastructure, water resources, agriculture, disaster risk reduction and coastal zone management.

(d) Outcomes and reports of completed multilaterally and bilaterally funded projects within the region, for example the Global Climate Observing System Regional Workshop Programme (GCOS RWP), Assessment of Impacts and Adaptation to Climate Change (AIACC), Advancing Capacity to Support Climate Change Adaptation (ACCCA), and other relevant documents.

(e) The Global Environmental Outlook (GEO3, GEO4) report and database on the Asia Region.

(f) Examples of Community Based Adaptation Projects and Activities, including those in the UNFCCC database on local coping strategies as well as those presented in the CBA Workshops held in Dhaka in 2006 and 2007.

CONTEXT FOR ADAPTATION TO CLIMATE CHANGE

Asia is the world's largest and most populous continent which covers about 30% of the world land area and contains more than 60% of the current human population. The Asian continent is bounded on the north by the Arctic Ocean, on the east by the Pacific Ocean, and on the south by the Indian Ocean; the western boundary with Europe, runs roughly north-south along the eastern Ural Mountains, the Zhem River, the Caspian Sea, the Kuma-Manych Depression, the Black Sea, the Aegean Sea, the Mediterranean Sea, the Suez Canal, and the Red Sea. The Tibetan Plateau, with an average elevation of more than 4,000 m, is the world's largest plateau and Mount Everest (8,848 m), lying near the southern border of this plateau, is the world's highest peak.

Asia comprises more than 4 billion people living in 46 different States with a large range of differing cultures, environments, historical ties and government systems. Asia's wealth differs widely within and between States. In terms of nominal GDP, the largest economies in Asia are Japan (4,622.8 billion US$) and China (1931.7 billion US$) (UNDP, 2006). However, in terms of Purchasing Power Parity, the People's Republic of China has the largest economy (7,642.3 billion US$) in Asia and the second largest in the world while Japan (3,737.3 billion US$), India (3,389.7 billion US$) and South Korea (985.6 billion US$) are the world's third, fourth and eleventh largest economies respectively (UNDP 2006).

In most nations, central governments have played a key role in economic planning to achieve development goals, and in many instances, have also formulated environmental policies. Progress has been made in overall poverty reduction in many Asian countries. The number of people living on less than US$1 a day dropped by nearly 250 million from 1990 to 2001. Sustained growth in China and the acceleration of India's economy contributed to such progress. Efforts at reducing malnutrition, however, have been less successful. The most severe problems are evident in South Asia, where almost half the children aged 0-5 years are malnourished.

Environmental protection is considered to be a fundamental element in achieving several Millenium Development Goals (MDGs) and is also a powerful engine for economic growth and poverty eradication. South Asia has increased access for many to safe sources of drinking water, with India contributing substantially to the overall positive trend. Another positive sign is the significant progress made in improving energy efficiency and providing access to clean technology and fuels in Eastern Asia and Southern Asia, although energy efficiency continues to decline in Southeast Asia.

The region still faces some formidable challenges in its effort to protect valuable natural resources and the environment. Land and ecosystems are being degraded, threatening to undermine food security. In addition, water and air quality are deteriorating while continued increases in consumption and associated waste have contributed to the exponential growth in the region's existing environmental problems. Furthermore, the region is highly subject to natural hazards. Notable examples include the 2004 Indian Ocean Tsunami, the 2005 Pakistan Earthquake, and the 2006 landslides in the Philippines. There is evidence of prominent increases in the intensity and/or frequency of many of the extreme weather events such as heat waves, tropical cyclones, prolonged dry spells, intense rainfall, tornadoes, snow avalanches, thunderstorms, and severe dust storms in the region throughout the1990s and beyond (IPCC-TAR). Impacts of such disasters range from hunger and susceptibility to disease, to loss of income and livelihoods, affecting the survival and human well-being of both present and future generations.

The environmental and social problems in Asia are not new; many of them are persistent and some are getting worse. The Brundtland report highlighted the risks twenty years ago, declaring that the loss of ecosystems, for example natural forests and wetlands, the degradation of water and air quality, and ever increasing consumption and associated waste generation have made people and ecosystems more vulnerable. Since 2000, Asia's

GDP growth has surpassed that suggested by the Brundtland Commission, but ecosystems and human health continue to deteriorate. The region faces many years of challenges in attempting to preserve the environment while alleviating poverty and improving living standards with limited natural resources.

Current Environmental and Socio-economic Vulnerabilities

Even though Asia is the largest continent on Earth and spreads over four climatic zones, it has environmental and socio-economic problems common to the entire region, in sectors such as water, agriculture and land use, ecosystem and biodiversity, and population. The majority of the problems in these sectors are related to management issues and are exacerbated by the associated impacts of climate change.

Water Sector: Balancing Water Use and Demand

Adequate water supply is one of the major challenges in Asia. The growing demand for water is driven by high population, increasing urbanization and the flourishing of economic and social development. Since Asian economies depend heavily on agriculture and hence on irrigation, this sector continues to be the most demanding on available water resources. Excessive withdrawals from surface waters and underground aquifers, pollution of freshwater resources by the industrial sectors and inefficient use of freshwater are also major causes of water stress. Climate change has the potential to exacerbate water resource stresses in many countries of Asia (IPCC-TAR).

The Arid characteristic of Middle Eastern countries makes water scarcity a major problem. Extensive irrigation systems allow for agricultural production in this region, although this can increase salinity in groundwater aquifers. Combined with decrease in precipitation due to climate change, the Middle East faces a daunting task of ensuring fresh water supply for its sustainability. For example, in Iran, agricultural water usage is about 81.4 billion cubic metres, which accounts for 94% of the total consumption. Drinking water accounts for about 5% of the overall consumption and the industrial sector for only 1% of the whole. The required water is primarily extracted from ground water resources, which supply 55% of water used. The rest is sourced from surface water using dams, irrigation and traditional methods (National Communication, Iran, 2003).

Parts of Asia suffer from both water abundance as well as water scarcity. Southwest Bangladesh faces extreme water shortages as well as acute moisture stress during the dry months, adversely affecting both the ecology and agricultural production. Also, floods during the monsoon season inundate an average of 20.5% of the country and can flood as much as 70% (Mirza, 2002).

The rate of retreat of glaciers in the Himalayan highlands and permafrost thawing in north Asia has rapidly increased in the last decade. The Himalayan glaciers feed seven of Asia's great rivers: Ganga, Indus, Brahmaputra, Salween, Mekong, Yangtze and Huang He, and ensure a year-round water supply to billions people.

Human activities, such as land-use change, water storage, inter-basin transfers, and irrigation and drainage, are modifying the hydrological cycle in many river basins. A change in the onset, continuity and withdrawal patterns of the summer monsoon in recent years has also led to considerable spatial and temporal variations in rainwater availability in parts of the continent (Lal, 2005). Furthermore, watershed quality over the past few years has been threatened by climatic variability and natural disasters. The impact extends to damage of sanitation facilities and contaminated groundwater (UNEP, 2005).

There have been some attempts and actions to deal with water in Asia. Major structural, technological, social and policy changes, including effective water-related policies, lifestyle changes and technological transfer, are needed to decrease water withdrawals and hence water stress in most Asian countries. In addition, improvements in water-use efficiency, especially in the irrigation sector, will have immediate positive impacts on water availability under warmer climate.

Pressures on Ecosystems

Over the last two decades enormous pressures have been put on Asia's ecosystems to support the ever growing demand for natural resources. The most affected ecosystems are coastal and marine, forests and mountainous regions. Coastal ecosystems play an important role in the region. They are the locus of land-ocean interaction. The region has a long coastline and over half of Asia's more than 4 billion inhabitants live on or near the coast and depend directly on coastal resources, such as mangroves and coral reefs, for at least part of their livelihoods (Middleton, 1999).

Mangroves are vital to coastal ecosystems and fulfill multiple functions, such as providing forest products and habitat for fish and shellfish; protecting coastal land from erosion; enhancing biodiversity; and protection from inundation, for example the large mangrove forests in Phang Nga, Thailand, significantly mitigated the impact of the 2004 Indian Ocean Tsunami (UNEP, 2005).The destruction and reduction of ecosystem services and functions, in turn reduce their contribution to human well-being. Deforestation has caused the rapid reduction of timber production, especially of the timber only found in natural forests, thus affecting livelihoods (SEPA, 2004).

Heavy reliance on marine resources across the region has resulted in the degradation of many coral reefs, particularly those near major population centres. Moreover, the higher sea surface temperatures have led to severe bleaching of the corals in coastal regions of Asia. Approximately 60% of its coral reefs are estimated to be at risk and mining and destructive fishing practices are the greatest threats (UNESCAP, 2005).

Sustainable Agricultural Land Uses

Land degradation is a serious problem in Asia because of the threat to food security, human livelihoods and overall sustainability of the region. Problems include salinisation due to poor irrigation practices, soil erosion, desertification and overuse of water.

Since agriculture is the main land use in Asia, land conservation as a tool of sustainable agriculture has been heavily emphasized. It can promote rural development, as well as increase food security and ecosystem vitality. Immediate responses are reforestation, redefining protected areas and adopting integrated approaches such as Integrated Watershed Management. Good governance is the basic foundation of any land conservation strategy. Besides providing appropriate legal and policy mechanisms for administering land ownership, it can foster the active participation of civil society in land reform efforts and ensure equitable distribution of the benefits from agrarian development. Many farmers in Asia are women. Their contribution, however, tends to go unnoticed. Effective land management and conservation schemes take into consideration the need to recognize and protect the rights of female participants in agricultural areas and the benefits derived from those areas (FAO, 2003).

General Climate Change Profile for Asia

Different Climatic Zones and their Characteristics

The six geographical sub-regions of Asia primarily fall under the four climatic sub regions as categories by IPCC-TAR with it distinctive characteristics:

Boreal Asia is located on the north of the continent between 50°N and the Arctic Circle. Boreal forests cover most of this region. Siberia, Russia, is the coldest region of the northern hemisphere in winter. Three Siberian rivers, the Ob, the Yenissey, and the Lena, contribute about 42% of total runoff from all rivers of the Arctic basin. Lake Baikal, in the heart of this region is one of the world's largest and oldest lakes and contains as much as 85% of Russian surface freshwaters.

Arid and semi-arid Asia: extends from 22°N to 50°N and from 30°E to 105°E and includes more than 20 countries of the Middle East and central Asia, many of which are landlocked. Many storms develop locally, and the maximum frequency of cyclones occurs in January and March, although very little precipitation is recorded over most of the region because of the lack of moisture. Grasslands, rangelands, and deserts dominate most of arid and semi-arid Asia.

Temperate Asia extends from 22°N to 50°N and from 105°E to 150°E and includes eastern China, the Japanese islands, the Korean peninsula and Mongolia. Much of the natural forest in the region has long been destroyed. Broad plains have been cultivated and irrigated, and natural grasslands have been used for animal husbandry for centuries.

Tropical Asia extends from 10°S to 28°N and from 50°E to 150°E and includes several countries of south Asia, which are influenced predominantly by the monsoons. The region is physiographically diverse and rich in biodiversity. Agriculture is the main industry in several countries. Exploitation of natural resources associated with rapid urbanization, industrialization, and economic development has led to increasing air and water pollution, land degradation, and other environmental problems in countries of this region.

According to the IPCC-TAR there are large variations in observed changes in temperature and rainfall across Asia.

Boreal Asia: There has been an observed average annual mean increase in surface air temperature of about 2.9°C in the past 100 years, which is more pronounced in the winter although summer temperatures in central Siberia have exhibited decreasing trends. Annual mean precipitation in Russia is decreasing about -4.1 mm/month/100 years, particularly during warm years. However, during the past 10-15 years, precipitation has increased, mostly during the summer-autumn period. Consequently, the levels of major aquifers have risen by 50-100 cm; the growth of groundwater storage has resulted in increasing ground river recharge and considerable low-water runoff.

Arid and Semi-arid Asia: In most of the Middle East, mean annual surface temperature has risen by about 1.3°C during 1894-1997; in the arid regions of China, air temperature has increased since the 1970s. In Pakistan, annual mean surface temperature has a consistent rising trend since the beginning of 20th century. Annual mean rainfall is considerably low in most parts of the arid and semi-arid region and temporal variability is quite high.

Temperate Asia: The average annual mean surface temperature in Mongolia has increased by approximately 0.7°C over the past 50 years; consequently, noticeable changes have taken place in the length of the cold and warm seasons. Surface temperature in northeast China has increased in winter but decreased in summer; in southeast China surface temperature has decreased 1-2°C. In Japan, the surface air temperature has shown a warming trend during the past century. The East Asian monsoon greatly influences temporal and spatial variations in rainfall.

Annual mean rainfall in Mongolia is 100-400 mm and is confined mainly to summer. Summer rainfall seems to have declined over the period 1970-1990 in Gobi; the number of days with relatively heavy rainfall events has dropped significantly. In China, annual precipitation has been decreasing continuously since 1965. The summer monsoon is reported to be stronger in northern China during globally warmer years although drier conditions have prevailed during globally colder years.

Tropical Asia: Several countries have reported increasing surface temperatures. In Vietnam, annual mean surface temperature has increased approximately 0.32°C over the past three decades and in Sri Lanka a gradually increasing trend is suggested of about 0.30°C per 100 years.

The warming trend over India has been reported to be about 0.57°C per 100 years. Hills and mountain ranges cause striking spatial variations in rainfall. In India, long-term time series of summer monsoon rainfall have no discernible trends, but decadal departures are found above and below the long time average alternatively for three consecutive decades. Recent decades have exhibited an increase in extreme rainfall events over northwest India during the summer monsoon and the number of rainy days during the monsoon along the east coast has declined.

Extreme weather events such as heat waves associated with extreme temperatures, extra-tropical and tropical cyclones, prolonged dry spells, intense rainfall, tornadoes, snow avalanches, thunderstorms, and dust storms are known to cause adverse effects in widely separated areas of Asia. There is some evidence of increases in the intensity or frequency of some of these extreme weather events on regional scales throughout the 20th century, although data analyses are relatively poor and not comprehensive (Balling and Idso, 1990; Bouchard, 1990; Agee, 1991; Yu and Neil, 1991; Ostby, 1993; Bardin, 1994; Born, 1996). There are also reports of an increase in thunderstorms over the land regions of tropical Asia (Karl et al., 1995). The frequency and severity of wildfires in grasslands and rangelands in arid and semi-arid Asia have increased in recent decades (IPCC-TAR).

Permanent glaciers in mountainous areas have vacated large areas during the past few decades, resulting in increases in glacial runoff and Glacier Lake Outburst Floods (GLOFs) and an increased frequency of events such as floods, mudflows and avalanches affecting human settlements. However, as mountain glaciers continue to disappear over the next decades, the volume of summer runoff will be reduced as a result of loss of ice resources with serious consequences for downstream agriculture. For example, low- and mid-lying parts of Central Asia are likely to change gradually into more arid, interior deserts, and lower Himalayan Countries will have more intense and frequent drought.

In recent years, GLOFs have happened in Nepal, India, Pakistan, Bhutan, and China in the Himalayan region. The most significant GLOF event since detailed recording started occurred in Dig Tsho Lake at Bhote Kosi in Nepal in 1985. It destroyed a hydropower project, numerous foot-bridges, and trekking trails and caused loss of many lives. Millions of hectares of protected areas and many human settlements are located in the downstream of potentially dangerous glacial lakes. GLOF events have widespread impacts on the region's socio-economic systems, hydrology, and ecosystems.

Temperate Asia has experienced many droughts in the 20th century. In China, droughts in 1972, 1978, and 1997 have been recorded as the most serious and extensive. In Japan, drought disasters are significantly more frequent during years following ENSO warm events than in normal years. Floods are also a threat, for example a large number of severe floods have occurred in China, predominately over the middle and lower basins of the Yangtze (Changjiang), Huanghe, Huaihe, and Haihe Rivers (Ji et al., 1993) and severe flooding with daily rainfall exceeding 25 cm struck during July and August 1998 in South Korea.

In Tropical Asia, droughts, floods and cyclones are the key natural disasters. In India, chronically drought-affected areas cover the western parts of Rajasthan and the Kutch region of Gujarat. Drought conditions have also been reported in Bihar and Orissa States. In Bangladesh, about 2.7 Mha are vulnerable to drought annually (Mirza, 1998). Drought or near-drought conditions also occur in parts of Nepal, Papua New Guinea, and Indonesia, especially during El Niño years. In India, Laos, the Philippines, and Vietnam, drought disasters are more frequent during years following ENSO events. At least half of the severe failures of the Indian summer monsoon since 1871 have occurred during El Niño years (Webster et al., 1998).

Floods are also common in this region. For example in Bangladesh the total flood-prone area is 3.1 Mha and in India it is about 40 Mha (Mirza and Ericksen, 1996). In the event of enhanced anomalous warming of the Eastern Equatorial Pacific Ocean, such as that observed during the 1998 El Niño, a higher frequency of intense extreme events across Asia is possible.

There is general agreement among global climate models (GCMs) of a significant acceleration of warming in the 21st century in Asia over that observed in the 20th century. Predictions indicate that higher latitudes in Asia will warm more than the global average and will receive disproportionately more precipitation, while mid-continental mid-latitude areas may become drier, depending on the effects of aerosols (IPCC-TAR; Gao et al., 2003).

The area-averaged annual mean warming is projected to be about 3°C by the 2050s and about 5°C by the 2080s over the land regions of Asia as a result of future increases in atmospheric concentration of greenhouse gases. The rise in surface air temperature is projected to be most pronounced over North Asia in all seasons. In general, projected warming over all sub-regions of Asia is higher during northern hemispheric winter than during summer for all time periods.

An increase in annual precipitation in most of Asia during the 21st century is predicted with the relative increase being largest and most consistent in North and East Asia (Lal, 2003; Kwon et al., 2004; Boo et al., 2005; Japan Meteorological Agency, 2005; Kurihara et al., 2005). There is a predicted decline in summer precipitation over the central parts of arid and semi-arid Asia, leading to expansion of deserts and periodic severe water stress conditions. Increased rainfall intensity, particularly during the summer monsoon, could increase flood-prone areas in East Asia, South Asia and Southeast Asia.

Future Change in Climate and Extreme Events

While there is no evidence that tropical cyclone frequency may change (Trenberth, 2005), an increase of 10-20% in tropical cyclone intensities (including the near-storm precipitation rates and destructive potential) for a rise in sea surface temperature of 2 to 4oC relative to the current threshold temperature is likewise projected in East Asia, Southeast Asia, and South Asia (Knutson and Tuleya, 2004; Emanuel, 2005).

Amplification in storm surge heights could result from the occurrence of stronger winds with increase in sea surface temperatures and low pressures associated with tropical storms resulting in an enhanced risk of coastal disasters along the coastal regions of East, South and Southeast Asian countries. The impacts of increase in cyclone intensities in any location will be determined by any shift in the cyclone tracks (Kelly and Adger, 2000).

Global warming has accelerated melting of glaciers, with the vast Himalayan glaciers showing the fastest rate of retreat. Recent studies by the International Commission of Snow and Ice (ICSI) and the DFID, have predicted that Himalayan glaciers might disappear within 40 years time causing water stress to millions in India and China depending on glacier fed rivers. The Chinese Academy of Sciences states that as many as 64 % of China's glaciers will not exist by 2050 if current trends continue.

Future changes in sea level rise are predicted to be geographically variable. An additional 0.5m sea level rise is projected for the Arctic during this century (ACIA, 2004). The rising rates of sea level vary considerably from 1.5 to 4.4 mm per annum along the East Asia coast due to regional variation in land surface movement (Mimura et al., 2004).

The projected rise of mean high water level is greater than that of mean sea level (Chen, 1991; Zhang et al., 2000). Thus projected relative sea level rise (RSLR), including that due to thermal expansion, tectonic movement, ground subsidence and the trend of rising river water level, is 40-60cm, 50-70 cm and 70-90 cm in the Zhujiang, Changjiang and Huanghe Deltas, respectively by the year 2050 (Li et al., 2004a, b).

Almost all Asian Counties have submitted their NC to the UNFCCC to comply with their commitments under the Convention. Three Asian countries, Bangladesh, Bhutan and Cambodia, have also submitted their National Adaptation Programmes of Action (NAPAs). Other LDCs in Asia are finalizing their NAPAs. The levels of assessment of impacts, vulnerabilities and adaptation to climate change and reporting varies widely by region and countries due to differences in institutional capacity and economic strengths.

Some countries have reported that adaptation measures have already been adopted to address climate change related problems. For example, the Law on Water Resources of the People's Republic of China has been promulgated to safeguard and normalize the management of water resources. The Project of South-to-North Water Diversion has been launched in order to increase water supply to Beijing, Tianjin and the Shandong Peninsula. The construction of upgrading projects for water-saving in 226 large irrigated areas and 200 demonstration projects for water-saving irrigation were launched to raise water use efficiency (Initial National Communication on Climate Change, China, 2004).

Terracing of slopes, one of the most effective amelioration activities, has been practiced in many mountainous and low mountain regions for a long time, for example Azerbaijan, Nepal, Indonesia and Bhutan. Besides prevention of erosion processes, terracing of slopes enables additional areas to be cultivated and improves land productivity. The NC of Azerbaijan reported that productivity of terraced slopes would increase by 8-10 times once the cultivation of fodder crops is practiced. Increase of terraced slopes' productivity is maintained through the improvement of plant water regime as a result of 2-3 fold decrease of surface drain.

National Programmes for Adaptation and Climate Change

Several NCs give an emphasis on "no-regret" adaptation options, particularly for the water and agriculture sectors, in order to meet national development goals and promote sustainable development (National Communication of India, Saudi Arabia, Bangladesh, China etc). For example, China stated that they are promoting sustainable development and utilization of water resources; reinforcing adaptive capacity of the water resource system; and thus reducing the vulnerability of the water resource system to climate change. Iran suggests a number of methods for conserving water used in agriculture such as improving irrigation efficiency by lining existing earth irrigation canals, using pipes as water conduits and the reuse of drainage water.

The government of Saudi Arabia has implemented measures to protect the sustainability of water resources. These measures include execution of a long-term investigation programme to assess the availability of groundwater and surface water resources in different regions and supporting the drilling of thousands of wells, under the supervision of the Ministry of Agriculture, for domestic, agricultural, and industrial purposes.

Key biophysical vulnerabilities to climate change, variability and extreme events in Asia vary widely across the region due to differences in physical, social and economic circumstances. It is revealed from the NCs and NAPAs of Asian countries that the key biophysical vulnerability contexts are agriculture and food security, water resources, natural ecosystem, forestry and biodiversity, coastal zone and sea level rise, natural disasters and human health.

In the past few decades, production of rice, corn and wheat has declined in many parts of Asia due to increasing water stress arising partly from increasing temperature, increasing frequency of El Niño and reduction in the number of rainy days (Agarwal et al., 2000; Jin et al., 2001; Fischer et al., 2002; Tao et al., 2003a; Tao et al., 2004). In a study by the International Rice Research Institute, the yield of rice was observed to decrease by 10% for every 1°C rise in growing-season minimum temperature (Peng et al., 2004). A decline in potentially good agricultural land in East Asia and substantial increases in suitable areas and production potentials in currently cultivated land in Central Asia have also been reported (Fischer et al., 2002). Climate change could make it more difficult than it is already to step up agricultural production to meet growing demands in Asian developing countries.

Recent studies confirm and extend the findings reported in the IPCC-TAR that temperature and precipitation changes in future decades will modify, and often limit, direct CO2 effects on plants. For instance, high temperatures during flowering may lower CO2 effects by reducing grain number, size and quality (Caldwell et al., 2005; Baker, 2004; Thomas et al., 2003). Increased water demand under warming may also reduce CO2 effects. Rainfed wheat grown at 450 ppm CO2 showed yield increases up to 0.8°C warming, then declines beyond 1.5°C warming; additional irrigation was needed to counterbalance these negative effects (Xiao et al., 2005).

Recent studies suggest that substantial decreases in cereal production potential in Asia could be likely by the end of this century as a consequence of climate change. However, regional differences in the response of wheat, maize and rice yields to projected climate change are likely to be significant (Parry et al., 1999; Rosenzweig et al., 2001).

Crop simulation modelling studies based on future climate change scenarios indicate that substantial losses are likely in rainfed wheat in South and Southeast Asia (Fischer et al., 2002). In South Asia, the drop in yields in non-irrigated wheat and rice will be significant for a temperature increase greater than 2.5oC, incurring a loss in farm level net revenue of between 9% and 25% (Kumar and Parekh, 1998). The net cereal production in South Asian countries is projected to decline at least between 4 to 10% by the end of this century under the most conservative climate change scenario (Lal, 2005).

PROJECTED FUTURE IMPACTS AND VULNERABILITIES

Agriculture and Food Security

Studies reveal that a northward shift of agricultural zones is likely such that the dry steppe zone in the eastern part of Mongolia would push the forest-steppe to the north resulting in shrinking of the high mountainous and forest-steppe zones and expansion of the steppe and desert steppe (Tserendash et al., 2005). Studies also suggest that, by the middle of this century in northern China, the tri-planting boundary will likely shift by 500 km from Changjiang Valley to Yellow River basin, and double planting regions will move towards the existing single planting areas, while single planting areas will shrink by 23% (Wang, 2002). Suitable land and production potentials for cereals could marginally increase in the Russian Federation and in East Asia (Fischer et al., 2002).

More than 28 million ha in South and East Asia require substantial increases in irrigation for sustained productivity (FAO, 2003). Agricultural irrigation demand in arid and semi-arid regions of Asia is estimated to increase by at least 10% for an increase in temperature of 1oC (Liu, 2002; Fischer et al., 2002). The rainfed crops in the plains of north and northeast China could face water-related challenges in coming decades due to increases in water demands and soil-moisture deficit associated with projected decline in precipitation (Tao et al., 2003b).

The impact of climate change on Asian fisheries depends on the complicated food chains in the surrounding oceans, which are likely to be disturbed by climate change. Fisheries at higher elevations are likely to be adversely affected by lower availability of oxygen due to a rise in surface air temperatures. In the plains, the timing and amount of precipitation could also affect the migration of fish species from the river to the floodplains for spawning, dispersal, and growth (FAO, 2003). Future changes in ocean currents, sea level, sea water temperature, salinity, wind speed and direction, strength of upwelling, mixing layer thickness and predator response to climate change have the potential to substantially alter fish breeding habitats and food supply for fish and ultimately the abundance of fish populations in Asian waters (IPCC-TAR).

Water availability, in terms of temporal as well as spatial distribution, is expected to be highly vulnerable to anticipated climate change. Growing populations and the concentration of the population in urban areas will exert increasing pressures on water availability and water quality. As reported in IPCC (1998), surface runoff generally is expected to increase in the high latitudes and near the equator and decrease in the mid-latitudes under anticipated climate change scenarios.

Some areas of the Asian continent are expected to experience increases in water availability; other areas will have reduced water resources available. Surface runoff is projected to decrease drastically in arid and semi-arid Asia under climate change scenarios and would significantly affect the volume of water available for irrigation and other purposes. Sensitivity analysis of water resources in Kazakhstan to projected climate change scenarios indicates that surface runoff would be substantially reduced as a result of an increase in surface air temperature of 2°C accompanied by a 5-10% decline in precipitation during summer (Gruza et al., 1997). In temperate Asia, future changes in surface runoff would be highly spatially variable. An increase in surface runoff seems likely in Mongolia and northern China. Studies suggest that, on average, a 3°C increase in temperature coupled with a 10% increase in precipitation will increase river flows by approximately 15% in water abundant areas. An increase in temperature also accelerates snow melting, which increases river flows from January through March but decreases flows from April through June (Hanaki et al., 1998; Inoue and Yokoyama, 1998).

Water Resources

The perennial rivers originating in the high Himalayas receive water from snow and glaciers. Snow, ice, and glaciers in the region are approximately equivalent to about 1,400 km3 of ice. The contribution of snow to the runoff of major rivers in the eastern Himalayas is about 10% (Sharma, 1993) but more than 60% in the western Himalayas (Vohra, 1981). Because the melting season of snow coincides with the summer monsoon season, any intensification of the monsoon is likely to contribute to flood disasters in Himalayan catchments. Such impacts will be observed more in the western Himalayas compared to the eastern Himalayas because of the higher contribution of snowmelt runoff in the west (Sharma, 1997). An increase in surface runoff during autumn and a decrease in springtime surface runoff are projected in highland regions of south Asia. In addition, the increase in surface temperature will contribute to a rise in the snowline, which, in effect, reduces the capacity of the natural reservoir. This situation will increase the risk of flood in Nepal, Bangladesh, Pakistan, and north India during the wet season (Singh, 1998). No significant changes are projected for annual mean surface runoff in southeast Asia; although an increase during winter and a decrease during summer season is likely.

Reduction of annual surface runoff and increase in temperature will reduce total availability of water both in surface and groundwater in the western region of Asia. The impacts will be greater on irrigation for agriculture and water supply for the industrial and domestic sector. In Saudi Arabia, for example, the total water stress is equal to the total quantities of decrease in groundwater recharge and surface runoff.

As a consequence of global warming, the present distribution of species in high-elevation ecosystems is projected to shift to higher elevations, although the rates of vegetation change are expected to be slow and colonization success would be constrained by increased erosion and overland flows in the highly dissected and steep terrains of the Himalayan mountain range (IPCC-TAR).

There is growing concern that climate change may accelerate the damage to freshwater ecosystems such as lakes, marshes, and rivers. With a rise in temperature, a decrease in the amount of snowfall in the Lake Biwa catchment in Japan is projected-which might exacerbate the process of eutrophication (Fushimi, 2000a). Deterioration of lake water quality also is suggested in Kasumigaura Lake in eastern Japan (Fujimoto et al., 1995; Fukushima et al., 2000). The response of lakes and streams to climate change will involve complex interactions between the effects of climate on areal inputs, hydrology, catchments and in-lake processes.

Climate change is expected to affect the boundaries of forest types and areas, primary productivity, species populations and migration, the occurrence of pests and diseases, and forest regeneration. The increase in GHGs also affects species composition and the structure of ecosystems, which, in turn, affects ecosystem function (Schulze, 1994). The interaction between elevated CO2 and climate change plays an important role in the overall response of net primary productivity to climate change at elevated CO2 (Xiao et al., 1998).

Climate change will have a profound effect on the future distribution, productivity, and health of forests throughout Asia. Warming is expected to be particularly large at high latitudes and so climate change could have substantial impact on boreal forests (Dixon et al., 1996; IPCC, 1996; Krankina, 1997) such as decreasing permafrost areas, improving growing conditions, decreasing areas of disturbed stands and reducing ecosystem quality. Moreover, forest fire is expected to occur more frequently in boreal Asia as a result of increased mean temperature (Valendik, 1996; Lelyakin et al., 1997). Pest activity also could increase with a rise in temperature, depending on the age composition of the boreal forests (Alfiorov et al., 1998).

Studies on projected impacts of climate change suggest that northeast China may become deprived of conifer forests. Broad-leaved forests in east China may shift northward by approximately 3° of latitude. These results are based on a 2°C increase in annual mean temperature and a 20% increase in annual precipitation (Omasa et al., 1996; Tsunekawa et al., 1996).

With an increase in temperature of 2-3°C combined with reduced precipitation as projected for the future in the semi-arid and arid regions of Asia, grassland productivity is expected to decrease by as much as 40-90% (Smith et al., 1996). Approximately 70% of pastures are facing degradation, with dramatic decreases in fodder yield over recent decades in some parts of Mongolia (Khuldorj et al., 1998). Climate change is likely to represent an additional stress to rapid social change in many of Asia's rangelands.

Ecosystems and Biodiversity

There are complex interrelationships and feedbacks between human driving forces and impacts, on the one hand, and climate- and sea level-induced changes and effects on the other (IPCC, 1996). At the interface between ocean and terrestrial resources, coastal ecosystems undergo stress from competing multi-usage demands, while having to retain their functional diversity and resilience in the face of global environmental change (Bower and Turner, 1998). Much infrastructure and economic activity is located in coastal zones, which are vulnerable to sea level rise and more frequent weather events (e.g. cyclones, storm surges). Temperature increases could also adversely affect local flora and fauna of coastal areas, as well as the biological equilibrium of marine life.

Global mean sea level is projected to rise between 9 and 88 cm between 1990 and 2100, with a central value of 48 cm, for the full range of SRES scenarios (Church et al., 2001). The projected central value of 48 cm gives an average rate of sea-level rise of 2.2 to 4.4 times the observed rate over the 20th century. Even with drastic reductions in greenhouse gas emissions, sea level will continue to rise for centuries beyond 2100 because of the long response time of the global ocean system.

A major uncertainty is how the global mean sea-level rise will manifest itself on a regional scale. All models analysed by the IPCC-TAR show a very uneven spatial distribution of sea-level rise. Some regions show a sea-level rise substantially higher than the global average (in many cases more than twice the average) and others a sea-level fall (Church et al., 2001). However, the patterns produced by the different models are not similar in detail. This lack of similarity means that confidence in projections of regional sea-level changes is low, although it is clear that they are important with respect to impacts of sea-level rise on coastal zones.

Climate change will have a wide range of health impacts all across Asia. Although a reduction in health stresses and wintertime deaths is anticipated as a result of less frequent occurrence of extreme cold temperatures in boreal and temperate Asia, an increase in the frequency and duration of severe heat waves and humid conditions during the summer will increase the risk of mortality and morbidity, principally in older age groups and urban poor populations of temperate and tropical Asia (Epstein et al., 1995). In addition, human health is likely to suffer chronically from heat stress related damages (Bouchama et al., 1991; Ando, 1998).

Adverse health impacts result from the build-up of high concentrations of air pollutants such as nitrogen dioxide (NO2), ozone, and air-borne particulates in large urban areas. Combined exposures to higher temperatures and air pollutants appear to be critical risk factors for health during the summer months (Piver et al., 1999).

Global warming will also increase the incidence of some diseases, such as respiratory and cardiovascular diseases, in arid and semi-arid, temperate, and tropical Asia. There is already evidence of widespread damage to human health by urban air quality and enhanced climate variability in Asia. Throughout newly industrialized areas in Asia, such as Chongqing, China and Jakarta, Indonesia, air quality has deteriorated significantly and contributed to widespread heat stress and smog induced cardiovascular and respiratory illnesses in the region (WHO, 2004).

Coastal Zones

Human Health

Malaria is one of the most important diseases in tropical Asia. With a rise in surface temperature and changes in rainfall patterns, the distribution of vectors such as mosquito species may change (Patz and Martens, 1996; Reiter, 1998) and temperature can directly influence the breeding of malaria protozoa and suitable climate conditions can intensify the invasiveness of mosquito (Tong et al., 2000). Changes in environmental temperature and precipitation could expand vector-borne diseases into temperate and arid Asia. The spread of vector-borne diseases into more northern latitudes may pose a serious threat to human health. Climate change is likely to have principal impacts on epidemics of malaria, dengue, and other vector-borne diseases in Asia (Martens et al., 1999). The epidemic areas of vector-borne diseases in Asia would depend on many demographic and societal factors, as well as environmental hygiene for vector control, available health infrastructure, and medical facilities. An empirical model projected that the population at risk of dengue fever (the estimated risk of dengue transmission is greater than 50%) will be largest in India and China (Hales et al.., 2002).

The global burden of climate change-attributable diarrhoea and malnutrition are already the largest in the world in Southeast Asian countries including Bangladesh, Bhutan, India, Maldives, Myanmar and Nepal in 2000, and the relative risks for these conditions for 2030 is expected to increase (Patz et al., 2005), although in some areas such as southern states in India there will be a reduction in the transmission season by 2080 (Mitra et al., 2004).

Warmer sea surface temperatures along coastlines of South and Southeast Asia due to climate change will support higher phytoplankton blooms. These phytoplankton blooms are excellent habitats for the survival and spread of infectious bacterial diseases such as cholera (Pascual et al., 2002). Water-borne diseases including cholera and a suite of diarrhoeal diseases caused by organisms such as giardia, salmonella and cryptosporidium could also become common with the contamination of drinking water quality.

The impacts of climate change on Asia will place additional stress on socioeconomic and physical systems. These pressures may induce change in demographic processes. Demographic trends, including the stability and size of populations, will be influenced directly through the impacts of climate change on human health and indirectly through the impacts of climate change on food security and the viability of natural resource-based economic activity. A further demographic response will come about through the risk of extreme events on human settlements. If the incidence and magnitude of events such as droughts and coastal floods increase, there could be large-scale demographic responses-for example through migration.

The annual rate of growth in migration on a global scale has been greatest in developing countries of south and southeast Asia. Migration in itself is not necessarily a signal of vulnerability to present-day extreme events. Motivations for migration are diverse; much rural to urban migration in Asia takes place as a result of increased economic opportunities in megacities (cities with at least 8 million inhabitants). For instance, population growth and land scarcity has encouraged the migration of more than 10 million Bangladesh natives to neighbouring Indian states during the past two decades. This migration has been exacerbated by a series of floods and droughts affecting the livelihoods of landless and poor farmers in this region. Land loss in coastal areas resulting from inundation from sea-level rise as a result of climate change is likely to lead to increased displacement of resident populations. Many South Asian countries increasingly expect the number of internally displaced persons to rise in future.

Immigrant labour often benefits both the donor and the host cities/countries (Connell and Conway, 2000). However, perceptions of regional/national identity, language/cultural differences, and fears of unemployment may contribute to increased hostilities between immigrants and nationals in years to come. Climate change will act in parallel with a complex array of social, cultural, and economic motivations for and impacts of migration (Pebley, 1998; Conway et al., 2000; Kates, 2000). Irrespective of resource constraints, developing countries in Asia have to better equip themselves through appropriate public education and awareness programmes with disaster preparedness measures, including infrastructure for effective resettlement of displaced people as a consequence of weather calamities.

CAPACITY FOR NATIONAL CLIMATE CHANGE STUDIES AND THE USE OF THEIR RESULTS AND CLIMATE DATA

Development and Human Dimensions

The UNFCCC has recognized the importance of research and systematic observation to reduce uncertainties regarding the effects of climate change and impacts and responses to it, particularly adaptation to climate change. Various decisions of the COP to the UNFCCC have been aimed at strengthening systematic observations, improving the exchange of climate data, supporting capacity-building, and encouraging the development of national plans. It has also recognized that data at sub-regional and national levels is essential to facilitate vulnerability and impact assessments and to provide ground truth for model outputs. There is a wide variation in national capacity to collect, process, and make available observational data, as well as use data for impacts and vulnerability assessment.

A wide range of data is being collected and used for developing future climate change scenarios, impacts, and vulnerability assessment in Asia, including atmospheric, oceanic and terrestrial data and information. In most countries, the meteorological or hydro-meteorological department of the government is responsible for collecting, processing and supplying data as well as maintaining infrastructure of the systematic observation system.

The number of systematic observation systems, types of equipment, capacity of computers and models, and number of trained and skilled people vary widely across Asia. Variation of economic strength and national priority have resulted in limited access to modern observational technologies and computerized systems for many developing and least developed countries. The models that were used to predict climate change variations, extreme events and its associated impacts were designed for the developed countries and need accurate baseline data to run efficiently. A major problem is the lack of reliable and accurate data, discrepancies in data and non-systematic data collection, especially for rainfall and temperature. Failure of equipment is a very common concern and time series data is often incomplete.

Some highly developed economies in Asia such as Japan and South Korea have invested both financially and technologically in climate change research and climate observation systems. Oil exporting countries like Saudi Arabia and rapidly developing economies like China have also realized the importance of climate change on development and the economy. They have invested in state of the art observation systems and technological knowledge. China has developed its own climate models and has used those, with the IPCC models, to assess vulnerability of the country to climate change.

Across Asia, variations in climate are heavily influenced by changes in oceanic conditions in the Indian Ocean and the Pacific. Monitoring trends of Sea Surface Temperature (SST) and Sea Level are therefore essential in order to assess their impacts, particularly important in view of the threat posed by cyclone/typhoon and rising sea level.

In addition to global networks, a number of important regional oceanographic programmes and experiments are underway in the waters of South and Southwest Asia. These programmes currently measure and report a range of oceanographic and atmospheric variables including SST, wave height, air pressure, air temperature, and surface winds. Nevertheless, a major deficiency has been the absence of a unified approach for the systematic and sustained observation of the unique and complex ocean-atmosphere system in the region.

The SAARC Meteorological Research Centre (SMRC) is a regional organization in South Asia that concentrates primarily on the research aspects of weather forecasting and monitoring. The research areas include weather prediction and compiling climatological information. In addition to monitoring weather phenomena, SMRC is also engaged in developing a networking system among the Member States.

Characteristics of Observation Systems and Quality of Data

Systematic Observation

There is general acceptance that strengthening and upgrading the observation networks including telecommunications, data management, and smoothing out the irregular distribution of stations is necessary.

In Asia there are many countries facing specific problems such as delays in the maintenance and replacement of equipment, resulting in data gaps. Equipment that cannot be repaired is sometimes not replaced resulting in datasets that cannot be compiled and compared.

Decision 5/CP.5 of the UNFCCC COP invited the Global Climate Observing System (GCOS) to launch a regional workshop programme to facilitate improvements in climate observing systems. GCOS has organized three regional workshops and prepared regional action plans for the regions of East and Southeast Asia, South and Southwest Asia, and Central Asia.

All regional workshops have identified several global observational networks for the atmospheric component of the GCOS Initial Observing System (IOS). Specifically, a GCOS Surface Network (GSN) has been defined, a geographically representative GCOS Upper Air Network (GUAN) has been specified, and the Global Atmosphere Watch (GAW) network is now considered a component of the GCOS.

The GCOS workshops also recognized that these baseline networks represent the minimum required for characterizing global climate and also represent a stable and sustainable underpinning for national networks that operate on finer temporal and spatial scales. It is particularly important that stations in these global networks meet GCOS standards, operate continuously, produce high quality observations and deliver data and associated metadata in a timely fashion to designated GCOS data processing and archiving centers.

The recommendations and projects in this GCOS Action Plan are grouped under three overarching themes or planning elements: (a) Needs assessment, national reporting and preparation of national plans to meet user requirements for climate system data, (b) Capacity-building and infrastructure investment for improved planning and response to climate change, climate variability and climate extremes, and (c) Improving coordination at national and regional levels.

The Action Plans came up with eight projects for each region and enables: (a) ensuring sustainability of the existing network, and quality and data standard; (b) initiative for operational oceanographic network and increase representative observations; (c) improving hydrological network and quality of data; (d) increasing observations in large lakes and glaciers in Asia; (e) increasing applications of satellite and remote sensing for data generation and impact assessment.

The fourteenth session of the Steering Committee of GCOS held in 2006 asked for its advice on follow-up to the Regional Workshop Programme and potential for future actions in other areas of the developing world highlighted in the regional workshop programme. Most Asian countries have submitted their Initial NC. Some countries are now in the process of submitting their second NCs. Agriculture, water resources, ecosystem and biodiversity, coastal resources and human health are common sectors for which impacts and vulnerability assessment have been carried out using different models and scenarios. Adaptation options have been identified to reduce impacts and to help in meeting national development goals including poverty alleviation, livelihoods and disaster risk reduction.

For generating climate change scenarios, the NCs used General Circulation Models (GCMs) and Regional Circulation Models (RCMs), downsized using MAGICC/SCENGEN software. The most commonly used scenarios were those of the IPCC Special Report on Emissions Scenarios (SRES) for 'reference' and 'policy' scenarios. Some of the other models used to predict climate change impact on countries in Asia are CSIRO climate model and Hadley Centre Regional Climate Model (HadRM2).

Regional Action Plan to Improve Systematic Observation in Asia

Along with the models available globally and various scenarios published by the IPCC, a number of countries used additional tools and models for better impacts and vulnerability assessment at national and sub-national level. Furthermore analysis of historical data on temperature and precipitation was used to strengthen the findings of the future climate change trends.

Climate Scenario Generation, Impact Models and their Use

Resolution of the global scale models are not very suitable for assessing adaptation at local level although they provide scale of possible impacts and potential vulnerabilities of the biophysical system. Since the GCM model uses a very broad scale, it had to be downsized and used in conjunction with regional circulation models (RCM) to get better results. A major problem encountered when using models to get national level results was the need for readjustment to suit a country's individual needs.

National Adaptation Programmes of Action in Asia's LDCs are being prepared using guidelines developed by the COP and elaborated by the LDC Expert Group (LEG). Bangladesh, Bhutan and Cambodia have already submitted their NAPAs. Other LDCs are developing their NAPAs. The bottom-up approach of NAPAs allows for vulnerability and adaptation assessments relevant to national circumstances, particularly those relating to food-security, poverty alleviation and livelihood options. This approach results in the identification of immediate and urgent needs to deal with adverse impacts of climate change and variability.

Involvement of different stakeholders (national, sectoral, local) are an integral part in the preparation process for assessing impacts, vulnerabilities and adaptation measures, while maintaining urgency and immediacy as the main principle of the NAPA. For example, Bangladesh has involved different stakeholders at the different stages of its NAPA formulation, allowing facilitation of mainstreaming adaptation into national and sectoral development policies, strategies and programmes.

Vulnerability-Based and Other Approaches to Adaptation and their Use

The preparation process has engaged policy makers of Government, local representatives of the Government (Union Parishad Chairman and Members), scientific community members of the various research institutes, researchers, academicians, teachers (ranging from primary to tertiary levels), lawyers, doctors, ethnic groups, media, NGO and CBO representatives and indigenous women.

At the highest level it has a Project Steering Committee (PSC) headed by Secretary Ministry of Environment and Forest. The Project Steering Committee is represented by high level officials and experts from different government and non-government organizations to provide guidance. In addition to the Ministry of Environment and Forest, other noteworthy government ministries and agencies involved in the PSC are the Ministry of Planning, Economic Relation Division, Ministry of Agriculture, Ministry of Food and Disaster Management, Water Resource Planning Organization, Ministry of Fisheries and Livestock, Ministry of Land and Department of Environment. The Project Steering

Committee is represented by government, non-government ainternational research institutes including Bangladesh Institute of Development Studies (BIDS),Bangladesh Forestry Research Institute (BFRI), Bangladesh Centre for Advanced Studiesand IUCN Ba nd (BCAS) ngladesh.

The second group of stakeholders was a multi-disciplinary team of experts and sectoral working groups with critical roles and responsibilities to analyze vulnerability of the natural, economic and social systems and connect it with overall development of the country. The six Sectoral Working Groups (SWG) are a) Agriculture, Fisheries and Livestock coordinated by Bangladesh Agricultural Research Council (BARC), b) Forestry, Biodiversity and Land-use coordinated by IUCN, c) Water, Coastal Zone, Natural Disaster and Health coordinated by Water Resources Planning organization (WARPO), d) Livelihood, Gender, Local Governance and Food Security coordinated by Bangladesh Institute for Development Studies (BIDS), e) Industry and Infrastructure coordinated by Department of Environment (DoE), and f) Policies and Institutes coordinated by BCAS.

The third group of stakeholders involved in the preparation process was from local and regional (divisional towns) level including people from the local government, local level non-government organizations, farmers and women. The local level stakeholders were involved in the regional consultation workshops. Objectives of the regional stakeholder consultation workshops were a) identification and ranking of existing problems related to variability, extremes and climate change, b) Identification of existing coping mechanisms and measures, c) Suggestions for improvement of existing measures, d) Identification of new measures and ideas to address anticipated future change in intensity and extent of present problems.

The identified problems and suggestions have been incorporated in the sectoral analysis and future programme of action. In addition to the regional stakeholder consultation workshops each sectoral working group has organized meetings with different sectoral agencies for checking the possibility of integrating plans into sectoral policies.

It is revealed from NCs and other assessment studies carried out in different Asian countries that top-down, scenario based approaches and tools, including impact and integrated models, are strong in terms of biophysical aspects of impacts and certain types of dynamic interactions, but do not perform well in representing human interactions and local abilities to adapt. Vulnerability based bottom-up approaches involving different stakeholders are more attuned to the local circumstances, and can give a picture of local level institutions that deal with the problems and productivity contexts. This approach is better able to represent local options and constraints than scenario-driven studies.

Institutional Capacity and Need

Institutional strength, capability and availability of reliable data are key issues for generating reliable climate scenarios. In developing countries of Asia, climate observation centers are mostly state governed institutions and therefore limited scope exists for research due to the lack of requisite technology/equipment, experts in the field and appropriate funding.

The situation is somewhat changing with the emergence of the threat of climate change to the development and economies of countries. Governments are investing in climate change studies and trying to provide technical knowledge in this area. There has been progress from private research and academic institutions with appropriate expertise that carry out climate change studies and predict scenarios. Furthermore, international organizations are supporting governments and private institutions to carry out advanced climate change research and assess vulnerability and adaptation options.

There is still a lack of baseline information for understanding the complex interplay between and within natural and human systems and a considerable gap in information on likely changes in climate and human systems in different ecosystems and agro-climatic systems. National scale assessment in many countries based on models failed to incorporate human dimensions, particularly livelihood aspects and inter-sectoral relationships. Without such national assessments as a sound basis for designing and planning adaptation policies, strategies and programmes, decisions on adaptation will remain uncertain and will not lead to effective results from implementation.

Understanding the effects of climate change, and associated enhanced climate variability and sea level rise, at the local and national levels is critical for adaptation, as is the capacity to select and apply appropriate methods and tools to prepare for adaptation.

Furthermore, the development of technologies for adaptation; the use and integration of traditional knowledge for developing adaptation options; integration of local knowledge and scientific knowledge for technology transfer; and communication of vulnerability and adaptation both to vulnerable communities and policy makers are also essential.

Documentation of successful pilot schemes on adaptation assessment with emphasis on processes; development of sharing mechanisms like the adaptation database of the UNFCCCand stronger public education and awareness programmes are all useful in disseminating information relating to adaptation activities to climate change and its benefits.

There is a need for building awareness about the potential impacts among the concerned people, and developing good quality databases. Systematic efforts are required to study the impact assessments of different climatic parameters. Studies about future projections of changing regional climate provide insights for methodological developments, including models for integrated assessment and GIS-based computer algorithms for supporting policy assessments at regional levels.

Considering commonality of problems, varied capacity (economic, social, and institutional) and existing experiences on assessment of impacts, vulnerability and adaptation, relevant cooperation among Asian countries would be helpful. Sharing experiences on disaster

risk reduction and preparedness and integration of climate change into disaster risk reduction are also necessary.

ADAPTATION TO CLIMATE CHANGE

Adaptation Assessment and Work in the Context of the UNFCCC

The Kyoto Protocol created opportunity to participate in Clean Development Mechanism projects. Article 6 of the Convention also allows Parties to implement activities on education, awareness raising and training.

NCs and NAPAs of Asian countries have reported existing coping strategies and adaptation measures required to reduce vulnerability to climate change. Many adaptation measures reported to have been adopted are win-win options and also support national development goals and objectives.

Possible Adaptation Options

Despite differences in physical and socioeconomic characteristics, the countries of the Asian region have much in common as far as vulnerability to climate change is concerned. The sectors affected by climate change, variability and extreme weather events across Asia include agriculture, water resources, ecosystem and biodiversity and human health.

However, priority areas to deal with climate related problems vary due to the scale of impacts and national development priorities, for example agriculture and water in Tropical Asia; ecosystems and water in Arid and Semi Arid Asia; water and agriculture in Temperate Asia.

In order to address impacts of climate change on different sectors, a sectoral approach has been suggested. Due to the complex nature of climatic problems and sectoral inter-dependency, for example with agriculture and water, adaptation options in one sector will contribute to others. A list of the identified adaptation measures are given below.

- Adjust cropping calendar and crop rotation to deal with climatic variability and extremes
- Develop and promote use of high-yielding varieties and sustainable technological applications (drought and saline tolerant varieties)

Water Resources (flood and water scarcity)

- o Develop flood-and drought-control management systems
- o Reduce future developments in flood plains
- o Use appropriate measures for protection against soil erosion
- o Conserve groundwater supply, water impoundments, and efficient water resource systems
- o Improve preparedness for water related natural disasters

Ecosystems and Biodiversity

- o Introduce integrated ecosystem planning and management
- o Reduce habitat fragmentation and promote development of migration corridors and buffer zones
- o Encourage mixed-use strategies
- o Prevent deforestation and conserve natural habitats in climatic transition zones inhabited by genetic biodiversity of potential for ecosystem restoration

Coastal Resources

- o Protect wetlands and allow for migration
- o Prepare contingency plans for migration in response to sea level rise
- o Improve emergency preparedness for weather extremes (e.g. cyclones and storm surges)
- o Evaluate coastal subsidence rates in sensitive coastal regions for better coastal land use planning
- o Implement existing coastal zone management plans
- o Protect marine resources

Human Health

- o Build heat-resistant urban infrastructures and take additional measures to reduce air and water pollution
- o Adapt technological/engineering solutions to prevent vector-borne diseases/epidemics
- o Improve health care systems, including surveillance, monitoring, and information dissemination
- o Improve public education and literacy rate in various communities

- o Increase infrastructure for waste disposal
- o Improve sanitation facilities in developing countries
- o Increase share of the forest (afforestation and reforestation) and forest protection
- o Introduction of endangered species into similar habitats, which will survive in case of climate change
- o Preservation of gene material in seed banks
- o Prevention of desertification processes
- o Monitoring of vulnerable ecosystems for better planning and readjustment

Water Resources

- o Improvement of water resources management systems including regulation of river flows, increase systems efficiency
- o Increase of accumulation volumes of winter-spring river flow and construction of water reservoirs
- o Reconstruction of irrigation systems in order to reduce water loss and apply advanced water-saving methods of irrigation
- o Economic and rational use of water
- o Adjusting the operation of existing and planned water supply infrastructure
- o Reducing the extremely high demand for freshwater in the agricultural sector
- o Monitoring of water resources to readjust national and sectoral plans

Agriculture

- o Change in agriculture system (introduction of humidity-preserving technologies, application of advanced agro-technical measures, and the introduction of new frost resistant, low water use and drought-resistant high-yielding varieties)
- o Revision of the zone distribution of irrigated lands
- o Enhanced resistance to diseases and pests

- o Reconstruction of existing irrigation system (introduction of sprinkling and drip irrigation)
- o Awareness of agrarian public about expected consequences related to forecasted climate change and other measures
- o Development of salt tolerant varieties and increased efficiency of water use for agriculture sector

Ecosystems and Biodiversity

- o Afforestation of eroded mountain slopes, binding of sands, rehabilitation
- o Forest tree improvement and use of fast growth species for reforestation and forestation

Coastal Zone and Sea Level

- o Protection of settlements, industrial and infrastructure facilities in the Coastal Zone
- o Introduction of soil and water saving technologies
- o Provision of drainage congestion and irrigated lands
- o Improvement and re-cultivation of salinized soils
- o Integrate sea level rise considerations into national development policies and capacity-building for relevant agencies and department

Human Health

- o Social, sanitary, preventive and administrative measures
- o Tracking of water quality, water treatment efficiency, and soil quality
- o Establishing a modern water conservancy management system and strengthening the unified management and protection of water resources
- o Increasing the capacity of reservoirs and river dams (prevent floods, tapping water sources to increase water-supply capacity)
- o Enhancing the protection and building of ecosystem (restoring vegetation cover, preventing and controlling soil erosion and loss)
- o Collection of rainwater for domestic use

Agriculture

- o Building up a water-saving agriculture and industry (popularizing water-saving irrigation, developing sprinkle irrigation and dripping irrigation, extending the use of water-saving facilities so as to increase water use efficiency)
- o Adjustment of the structure of agricultural production (introduce early-maturing crops with short growing period, selection and cultivation of new varieties)
- o Improve land cultivation management
- o Studies on new crop varieties, cultivation of alternative crop species
- o Set appropriate sowing and planting dates according to expected temperature and precipitation

Forest and Biodiversity

- o Enhance forest management (controlling and stopping the deforestation and the ecological damage)
- o Implementation of policy for the protection of natural forests
- o Plan to develop forestry planting and growth technology for each respective species
- o Preventive measures for forest hazards

Coastal Resources

- o Changes in design and standard for coastal infrastructure (consideration of impacts of sea level)
- o Mangrove afforestation
- o Relocate basic structures and industries and establish harmony between environment and industry
- o Protective programme for coastal erosion and coastal facilities

Human Health

- o Reinforcement of disease prevention measures
- o Epidemic prediction programmes

- o Strengthen preventative medicine activities in order to cope with possible vector borne diseases both for humans and domestic animals

Natural Disasters

- o Strengthen the early warning system within the national meteorological and hydrological service
- o Undertake full assessment of wild fire-risk zones and increase public awareness

The NAPAs have identified immediate and urgent needs to deal with adverse impacts of climate change. For example, the Bangladesh NAPA gave emphasis on coastal afforestation, development for agriculture, water resource management, increasing infrastructure resilience community based adaptation, and insurance options. The Bhutan NAPA gave emphasis on disaster management, landslide restoration, artificial lowering of glacier lakes, and hazard zoning. Awareness raising and capacity-building for vulnerable communities, planners and policy makers are common in the NAPAs.

Enabling Activities and Integration of Adaptation

Future impacts, vulnerability and adaptation assessment can be built on initial NCs and an integration of scenario based and vulnerability based assessment would be helpful for assessing both vulnerability of biophysical and social systems. This would provide better assessment, design, planning and implementation of adaptation projects. Involvement of different policy makers (national and sectoral) and other stakeholder groups in the preparation process will give a sense of ownership and future integration into sectoral policies and plans.

The link between climate change and development is obvious. Climate change impacts will significantly affect national development, particularly amongst the world's poorest communities. Showing linkage between adaptation to climate change and its contribution to development, and how adaptation to climate change will bring increased sustainability is crucial for motivating national and sectoral policy makers.

Many local communities are already adapting to the impacts of climate variability and climate change on a daily basis. Their experiences can offer lessons for national governments wishing to support adaptation activities. Experiences from disaster preparedness, and mainstreaming climate change adaptation measures to disaster management are essential.

Existing Local and Sectoral Coping Strategies

There are very limited experiences available on adaptation to climate change, but the Asian region has a large body of knowledge and experience at the local level on coping with climatic variability and extreme weather events such as floods, drought and cyclones/ typhoons. Experiences of different sectors particularly infrastructural interventions like large scale irrigation for agriculture, coastal defence and flood protection, cyclone centre and warning systems are also significant in the Asian region.

Traditionally, farmers have observed a number of practices to adapt to climate variability, for example intercropping, mixed cropping, agro-forestry and animal husbandry. Moreover, over the years many Asian countries have adopted both surface water and groundwater irrigation and diversification in agriculture to deal with drought and structural or non-structural measures to deal with flood and coastal inundation.

Separate adaptation strategies for each country are not available but National Communications have identified and highlighted several adaptation needs and strategies for implementation. The most common criteria used in prioritizing adaptation to climate change are: complementarity to national development goals and objectives, cost-effectiveness, and building on existing practices. There is a need for capacity-building and awareness raising of targeted institutions and an emphasis on mainstreaming adaptation to climate change in national policy.

Several Community Based Adaptation (CBA) activities to climate change, variability and extreme events are implemented or being implemented in Asia. Most of these adaptation activities are small-scale and concentrate on agriculture, water and natural disaster amelioration. Most of the community based adaptation projects have an emphasis on livelihood of the impacted community, diversification of agriculture, conservation of water and awareness raising to change practices.

The NC of Saudi Arabia reported that it has already implemented a number of projects to deal with climate related problems. These include construction of 215 dams for water storage, installation of 30 desalination plants, enactment of water protection and conservation regulation, a leakage detection and control scheme, an advanced irrigation water conservation scheme and a system for modification of water pumping.

Traditional as well as technological approaches are used to cope with the risk of drought in India. Technological management of drought uses medium (seasonal) to long-term (annual to decadal) forecasts that are formulated using appropriate models. This information is then translated into early warning, and subsequently appropriate drought protection measures are taken.

In northern Sri Lanka, local farmers traditionally use Zero-tillage paddy cultivation which they say requires considerably less water than more common methods of rice planting. Over-grown weeds are slashed and removed from the paddy field, leaving the stubble intact. The paddy seeds are then dry-sown in the field over the stubble, and immediately covered with 8 cm of dry-paddy straws. The community elders use astrological consultations and religious ceremonies to determine an auspicious time and date to release the water. Farmers do not use any chemical fertilizers or pesticides. Average yields per hectare are about 20-30% less than with modern techniques, however profits are still relatively high, as the practice is relatively cheap.

In the Philippines, after Typhoon Sisang in 1987, which completely destroyed over 200,000 homes, the Department of Social Welfare and Development decided to instigate a programme of providing typhoon-resistant housing for those living in the most typhoon prone areas. The Core Shelter houses are designed to withstand wind speeds of 180 km/h and have typhoon resistant features.

Adaptation Strategies

The IPCC-TAR states that many early signs of climate change are observed and these are predicted to become more prominent over the 21st century. Adaptations are already required to deal with vulnerabilities associated with climate variability. Priority areas for adaptation are land and water resources, food productivity and security, and disaster preparedness and planning particularly for poorer, resource dependent countries.

For many developing countries in Asia, climate change is only one of a host of problems, including poverty, hunger, water supply, waste, pollution and energy. Considering the availability of resources for adaptation and that adaptation responses are closely linked to development activities, then development of adaptation options need to be evaluated along with present and future development plans. Strategies require local involvement, inclusion of community perceptions, and recognition of multiple stresses on sustainable management of resources.

Vulnerability to climate change is complex and multi-dimensional, and there is no silver bullet for adaptation. Therefore the best choice may be a function of many factors pertaining to economic efficiency, risk reduction, robustness, resilience, reliability, etc. A common Framework for developing adaptation strategies is essential at the country level in order to design and implement adaptation options in an integrated manner. Local level planning and management strategies can be evolved and validated through this framework, so as to generate and evaluate various options suitable for local conditions.

Adaptation is location specific and therefore implementation of adaptation measures at cross-national and regional level would be difficult. However, maladaptation can be avoided, particularly in large watersheds shared by more than one country or region. Most importantly, processes of identifying adaptation measures, development of policies and strategies, engagement of different stakeholders, and linking adaptation with development can be shared among the countries having common characteristics and strength of economy. For example, experiences of a desalinization plant in a coastal area and drip irrigation system for agriculture for increasing efficiency of water can be shared and synergized.

Institutional capacities are a key element of community-based adaptation. Not just within the communities themselves, but also the institutions that interface with them. External actors who have already built up trust over years of working with the community, need an intermediary with the capacity to understand the problem. For successful capacity-building and to tap into more funding, the private sector needs to be involved. However, the full amount of committed official development assistance has yet to be disbursed. Climate change will require additional resources, which need to be allocated as agreed.

OPPORTUNITIES FOR ADAPTATION

Under the Convention and its Kyoto Protocol three funds have been created to support adaptation to climate change. The Global Environment Facility (GEF) is operating the Special Climate Change Fund (SCCF) and the Least Developed Country Fund (LDCF). Institutional arrangements to operate the Adaptation Fund under Kyoto Protocol are under discussion and expected to be finalized by COP13.

COP guidance on GEF support for adaptation identified three stages. Stage I provided support for the National Communications process, a portion of which is the vulnerability and adaptation assessment. Stage II provides further assistance for other capacity-building efforts for adaptation. Stage III refers to support for actual adaptation activities, including insurance, and has been implemented in the form of the GEF Strategic Priority on Adaptation (SPA). The GEF has allocated US$ 50 million under SPA of which US$ 5 million is devoted to piloting community adaptation initiatives through the Small Grants Programme (SGP).

The goal of the Community-Based Adaptation (CBA) Programme is to pilot the community component of the GEF SPA, and provide the basis upon which the GEF and other stakeholders can effectively support small-scale adaptation activities. This goal will be realized through three immediate objectives:

(a) development of a framework, including new knowledge and capacity, that spans the local to the intergovernmental levels (cross-scale 'policy laboratories'), to respond to unique community-based adaptation needs;

(b) identification and financing of diverse community-based adaptation projects (small-scale 'policy laboratories') in a number of selected countries; and

(c) capture and dissemination of lessons learned at the community level to all stakeholders, including governments.

The Least Developed Countries Fund was established to support, inter alia, the preparation and implementation of NAPAs. The operational modalities and procedures have been finalized and one project for Bhutan has already been approved under this fund. At present the LDCF has approximately US$ 115 million for funding priority activities in 48 LDCs under the UNFCCC.

The Special Climate Change Fund was established to finance developing country activities in adaptation, technology transfer, key sectors (energy, transport, industry, agriculture, forestry and waste management), and economic diversification for countries with economies dependent on the fossil fuel sector.

The Adaptation Fund is intended to fund concrete adaptation projects and programmes in developing countries that are particularly vulnerable to the adverse effects of climate change. The funding is provided by a 2% levy on proceeds from Clean Development Mechanism (CDM) projects (excluding those undertaken in LDCs), and "other sources". The total scale of the Adaptation Fund will therefore depend on the volume of CERs purchased through the CDM and the market value of those CERs. Based on a World Bank estimate of current CDM projects in the pipeline and an assumed CER market value of $15/CER, the fund would hypothetically stand at $102 million at present. Low and high estimates (of 900 million tonnes of carbon equivalent (MtC) and 2000 MtC respectively) for the first Commitment Period volume of CERs yield estimates between $270 M to $600 M for the total scale of the Adaptation Fund.

The GEF, under its different funding streams, has supported enabling activities and projects to support adaptation to climate change in Asia. Adaptation projects are also being implemented by other bilateral agencies and non-government organizations in the Asia Region.

In addition to ongoing adaptation projects, there are also other opportunities at the regional scale. At least 14 major international river watersheds exist in Asia. An integrated and decentralized system of restoration and conservation of the water cycle in these drainage basins is vital to reduce the negative consequences of natural and externally imposed perturbations. Watershed management is particularly challenging in countries where the people-to-land ratio is high and policy and management are inadequate, prompting use of even the most fragile and unsuitable areas in the watersheds for residential, cultivation, and other intensive uses. Many watersheds suffer badly from deforestation, indiscriminate land conversion, excessive soil erosion, declining land productivity, erratic and unreliable surface and groundwater resources, and loss of biodiversity. In the absence of appropriate adaptation strategies, these watersheds are highly vulnerable to climate change. Global climate change may also have serious water management implications on the territory of boreal Asia.

REGIONAL AND INTERNATIONAL COLLABORATION

Considering the importance and need for capacity-building to deal with climate change, a number of networks are working in the Asia Region. Most of them are supporting research, generating knowledge, bringing new ideas and debate, and sharing knowledge among different climate communities including researchers, policy makers and negotiators. The Asia-Pacific Network for Global Change Research (APN) is an inter-governmental network whose primary purposes are to facilitate global environmental change research in the Asia-Pacific region, increase developing country participation in that research, and strengthen links between the science community and policy makers. It promotes, encourages and supports research activities on long-term global changes in climate, ocean and terrestrial systems, and on related physical, chemical, biological and socio-economic processes. It realized that global change research cannot succeed without a high level of international cooperation and thus supports collaborative projects implemented under international research programmes.

Projects being implemented by APN can be categorized as Capacity-building, Comprehensive Research, and Adaptation to Climate Change, for example the 'Linking Climate Change Adaptation to Sustainable Development in Southeast Asia' project. In Asian LDCs, the network has built capacity through transferring tools and methodologies, assessment, training and awareness raising. It has also generated new knowledge on climate change and its impacts on water resources and agriculture.

START, the global change SysTem for Analysis, Research and Training, provides an international framework for capacity-building in scientific knowledge and technology to conduct regional and local research and to inform and influence decision-makers. START has forged strong linkages with regional governmental organizations, such as the Asia-Pacific Network for Global Change Research (APN), the Inter-American Institute (IAI), the New Partnership for Africa's Development, and the European Commission (EC). START furthermore collaborates with other organizations dedicated to building scientific research capacity across disciplines in developing regions, such as the Third World Academy of Science (TWAS), the International Foundation for Science (IFS), and the United Nations University (UNU). The START Regional Networks include Southeast Asia, South Asia and Temperate East Asia.

START completed its Assessments of Impacts and Adaptation to Climate Change (AIACC) project in Asia where potential impacts of climate change and vulnerability and adaptation assessment have been carried out for some sectors and countries. Advancing Capacity to Support Climate Change Adaptation (ACCCA) is an ongoing project of START focusing on community-based adaptation measures to weather related disasters, integration of adaptation strategies into development policies by effectively communicating climate risks and adaptation measures in agriculture and water sectors and mainstreaming climate change adaptation in watershed management and upland farming. The International Center for Integrated Mountain Development (ICIMOD), with financial support from UNEP, is assisting governments with the development of an inventory of glacial lakes; establishment of early warning systems; and development of adaptation measures. Inventories have been completed for Bhutan, Nepal, Astor basin in Pakistan, Pumqu basin in China, and Tista basin in India. Twenty glacial lakes in Nepal and 24 in Bhutan were identified as potentially dangerous and were recommended for further detailed investigation.

The Centre for International Forest Research (CIFOR) is an international research and global knowledge institution committed to conserving forests and improving the livelihoods of people in the tropics. CIFOR's high impact research helps local communities and small farmers gain their rightful share of forest resources, while increasing the production and value of forest products. CIFOR's three research programmes address the needs of the rural poor as well as environmental concerns. The Environmental Services and Sustainable Use of Forests Programme oversees research on biodiversity, carbon, fires, watershed functions, and the sustainable management and harvesting of forest products; the Governance Programme examines the process of making and implementing decisions about the management of forests by people and organizations beyond the scale of the individual household or small enterprise; and the Forests and Livelihoods Programme closely investigates how forest resources and their management, use and trade contribute to the livelihoods of the rural and urban poor. The International Water Management Institute (IWMI) is a non-profit scientific organization funded by the Consultative Group on International Agricultural Research (CGIAR). IWMI's research agenda is organized around four priority themes covering key issues relating to land, water, livelihoods, health and environment. The Institute concentrates on water and related land management challenges faced by poor rural communities. The challenges are those that affect their nutrition, livelihoods and health, as well as the integrity of environmental services on which these depend. IWMI works through collaborative research with partners in the North and South, to develop tools and practices to help developing countries eradicate poverty and better manage their water and land resources. For disaster preparedness, rapid response and assessment, a Remote Sensed and GIS Based Knowledge Base System for Sri Lanka (KBS-LANKA) is being implemented by IWMI.

The Asian Disaster Preparedness Center (ADPC) is a non-profit organization supporting the advancement of safer communities and sustainable development, through implementing programmes and projects that reduce the impact of disasters upon countries and communities in Asia and the Pacific. ADPC is working on raising awareness, strengthening institutions, providing training and mainstreaming disaster reduction in development management in Asia. The Institute for Global Environmental Strategies (IGES) is a research institute that conducts pragmatic and innovative strategic policy research to support sustainable development in the Asia-Pacific region. IGES collaborates with a broad range of stakeholders, such as national governments, non governmental organisations, businesses and citizens groups, to carry out research, aiming to ensure that the results are reflected in the policy-making process. It has produced a number of policy documents on adaptation with special focus on adaptation beyond 2012. The World Conservation Union (IUCN) is the world's largest and most important conservation network working on climate change issues in Asia. This organization is involved in the National Capacity Self-Assessment Programme.

The Red Cross/Red Crescent Climate Centre aims to support the International Federation of Red Cross and Red Crescent Societies (IFRC) and other interested parties to reduce people's vulnerability to climate risks. The activities of the Climate Centre include:

(a) providing education and communication support about climate risk management within the Red Cross and Red Crescent Movement and among the general public;

(b) Supporting the integration of climate risk management into Red Cross/Red Crescent programmes through the 'Preparedness for climate change` programme that formally started on January 1, 2006;

(c) Bringing concerns about the impacts of climate change on vulnerable people and experiences with climate risk reduction to national and international policy makers;

(d) Analyzing, documenting and sharing knowledge and experiences on climate risk management. For example, the climate change group is working in Indonesia on integrated community based risk reduction and in Vietnam on community-based climate adaptation.

SouthSouthNorth (SSN) is a network-based non-profit organisation working in the fields of climate change and social development. They aim to reduce poverty by building capacity in the South and delivering community based mitigation and adaptation projects. SSN has developed a methodology for preparing community based adaptation projects known as SouthSouthNorth Adaptation Project Protocol (SSNAPP).

Capacity Strengthening of LDCs for Adaptation to Climate Change (CLACC) is an initiative of southern institutes working on sustainable development. Initiated in 2004 by the International Institute for Environment and Development (IIED) and the Regional and International Networking Group (RING) partner institutions, CLACC receives financial and technical support from a number of northern institutes. CLACC aims to strengthen the capacity of organizations working with civil society in environment and development while integrating climate change issues in the planning and implementation of projects. The Energy and Research Institute (TERI) in India develops solutions to global problems in the fields of energy, environment and current patterns of development which are largely unsustainable. The central element of TERI's philosophy has been its reliance on entrepreneurial skills to create benefits for society through the development and dissemination of intellectual property.

The institute not only identifies and articulates intellectual challenges straddling a number of disciplines of knowledge but also mounts research, training and demonstration projects leading to the development of specific problem-based advanced technologies that help carry benefits to society at large. TERI works closely with several multilateral organizations, national governments, corporate organizations, NGOs, and academia around the world. The Bangladesh Centre for Advanced Studies (BCAS) is an independent, non-profit, non-governmental, policy, research, and implementation institute working on sustainable development at local, national, regional and global levels. BCAS envisions promoting people-centred sustainable development by applying and advancing scientific, technical and local knowledge through research, by developing models, demonstration, policy advocacy and project implementation. It works with leading NGOs, research institutes, universities and government agencies in South Asia and other parts of the world.

SUMMARY AND ISSUES FOR CONSIDERATION

Key Observations

Asia still faces formidable challenges in its effort to protect valuable natural resources and the environment. Land and ecosystems are being degraded, threatening to undermine food security. In addition, water quality and availability are deteriorating while continued increases in consumption and associated waste have contributed to the exponential growth in the region's existing environmental problems. Furthermore, the region is highly subject to natural hazards. Countries in the Asian region have assessed impacts, vulnerability and adaptation to climate change using model-based and vulnerability-based approaches. The vulnerability-based or bottom-up approach was found to be good for identification of adaptation measures which are more people centred. The availability of data with the necessary geographic detail is the major limitation rather than computational capability or basic understanding particularly for agriculture, water and coastal zones.

Climate change induced key biophysical vulnerabilities vary widely across the region due to differences in physical, social and economic circumstances. It is revealed from the NCs and NAPAs that the most vulnerable sectors are agriculture and food security, water resources, natural ecosystems, forestry and biodiversity, coastal zones and sea level rise, natural disasters and human health.

The projected impacts of climate change are likely to exacerbate water stress and shortages in some regions and increase the frequency and intensity of floods and droughts. A wide range of precautionary measures are available at the regional and national level to reduce economic and social impacts of natural disasters. Integration of disaster risk reduction and climate change is emerging in many Asian countries. Adaptive capacities vary between countries, depending on social structure, culture, economic capacity, and the level of environmental disruptions. Limiting factors include poor resource and infrastructure bases, poverty and disparities in income, weak institutions, and limited technology. The challenge in Asia lies in identifying opportunities to facilitate sustainable development with strategies that make climate-sensitive sectors resilient to climate variability.

In general, financial, technological and institutional barriers usually hamper the implementation of adaptation measures to climate variability and change in many Asian countries, particularly LDCs. For example, the current water policy of India aims at integrated water resources development and management to tackle water stress, but its implementation is constrained by financial and technological limitations.

Future Needs

Many of the problems associated with climate impacts are also priorities on the development agenda and have been for a long time, irrespective of climate change. Environmental degradation and links between the environment and development have been well documented and analysed since the 1970s. Enough is known of the risks posed by climate variability and climate change to develop and implement adaptation projects immediately. Specific focus is necessary on the following issues;

(a) Development of effective adaptation strategies should take a more systems-oriented, people centred approach, emphasizing multiple interactive stresses, with less dependence on climate scenarios;

(b) Adaptation should aim to strengthen traditional coping mechanisms: optimizing current systems whilst building flexibility to cope with the uncertainties posed by climate change. Introducing new technology can be sustainable where it strengthens and builds on traditional approaches and reinforces local knowledge;

(c) An assessment of exiting coping strategies and its effectiveness under the warmer climate is necessary to modify practices for readjusting with changing condition that climate change will bring;

(d) Implementation of adaptation measures identified in National Communications and National Adaptation Programmes of Action should be supported effectively;

(e) Capacity-building and enhancement of technical capacity is needed to assess, plan and integrate adaptation needs into sectoral development plans that deal with the adverse impacts of climate change;

(f) The development of educational and public awareness programmes on climate change, the provision of public access to information on climate change issues and public participation (including NGOs) are important components of both the implementation of the UNFCCC and the development of national action plans;

(g) There is a need to support mainstreaming of adaptation into sectoral policy, particularly water, agriculture, coastal zones and managing natural ecosystems;

(h) More work is needed to promote mechanisms to increase funds for adaptation to climate change.

In order to address adaptation to climate change, including variability and extreme events, it is vital to continue to research the circumstances of the poor and vulnerable, both for the poor and vulnerable and on the poor and vulnerable. This requires stakeholder involvement, and will not be based exclusively on a top-down approach. This research must be communicated in a clear, targeted manner to relevant stakeholders including the communities themselves, NGOs active in the communities, local and national authorities, higher education institutes, global scientists, policymakers and decision makers.

2

Good Practices on Strategic Planning and Management of Water Resources in Asia and the Pacific

INTRODUCTION

Background

The following brief description of activities undertaken in the three phases of the implementation of the project on "Capacity-building in Strategic Planning and Management of Natural Resources in Asia and the Pacific" during the period from 2000 to 2004 illustrates the circumstances in which the good practices of the application of the strategic planning and management approach to the development of integrated water resources management (IWRM) plans were developed in the region.

a. Phase I of the application of strategic planning and management (SPM) into water resources management

Under phase I of the project, which took place in 2000 and 2001, efforts were made to review achievements in water resources management in the region, including those of ESCAP to form the basis for preparing a set of guidelines on strategic planning and management of water resources development. Important achievements made by the ESCAP secretariat identified in the review included the results of studies on "Integration of Water Resources Management into Economic and Social Development Plans in Asia and the Pacific" and "A Synthesis of Experience from the FAO-ESCAP Pilot Project on the Formulation of National Water Visions to Action".

These results together with international experiences and inputs from invited water resources experts in the region were compiled to form the first draft of the Guidelines on Strategic Planning and Management of Water Resources, which was discussed at the Regional Workshop to Finalize the Guidelines on Strategic Planning and Management of Natural Resources Development, held in Bangkok in December 2001.

At this regional workshop, the expert group also identified the overall framework for the introduction of SPM into the national development process taking into account the application of SPM at the national/basin, sectoral and organizational levels. In that context, the expert group identified priority purposes of the application of SPM, key elements or components of SPM, and initiation and implementation of SPM at these three levels. In order to ensure adequate preparation for the subsequent phases of the project, the expert group also identified priority follow-up activities at the subregional level as well as possible coordination of regional efforts on this subject.

b. Phase II of the application of strategic planning and management into water resources management

During phase II, the draft Guidelines was then used for training at five workshops for South-East Asia (July 2002), Central Asia and the Pacific (August 2002), South Asia (September 2002) and North-East Asia (October 2002). In addition, the draft Guidelines was posted on the website of ESCAP to encouraged interested water resources management experts to contribute to improving its applicability. In order to increase the applicability of the Guidelines for the preparation of integrated water resources management plans, several suggestions were solicited participants of the five subregional workshops to form the basis for the introduction of a new chapter on "Monitoring Outcomes of Strategic Planning in Water Management: Using Indicators Effectively" in the draft Guidelines, which was then posted on the ESCAP website in July 2003.

As part of the process on networking building on application of SPM to water resources management, the five subregional workshops were organized in cooperation with subregional counterparts as follows:

a. The subregional workshop for South-East Asia in July 2002 in cooperation with the Mekong River Commission Secretariat (MRCS) in Phnom Penh, Cambodia. The number of participants was 40, including 21 officials from eight countries of South-East Asia, namely Cambodia, Indonesia, the Lao People's Democratic Republic, Malaysia, Myanmar, Philippines, Thailand and Viet Nam, 19 experts from the Global Water Partnership (GWP), International Water Management Institute (IWMI), the MRCS and ESCAP.

b. The subregional workshop for the Pacific in August 2002 for both the energy and water sectors in cooperation with the South Pacific Applied Geoscience Commission (SOPAC) in Sigatoka, Fiji. 22 participants, including 18 officials from 12 countries of the Pacific, namely Cook Islands, Fiji, Kiribati, Marshall Islands, Nauru, Niue, Palau, Papua New Guinea, Samoa, Solomon Islands, Tonga and Tuvalu, three experts from SOPAC and an ESCAP Consultant participated in the Workshop on water resources.

c. The subregional workshop for Central Asia was held in cooperation with the Inter-State Commission on Water Management Coordination (ICWC) in Cholpan Ata, Kyrgyzstan, and was attended by totally 49 participants, including 25 officials of ICWC from the five countries of Central Asia, namely Kazakhstan, Kyrgyzstan, Tajikistan, Turkmenistan, and Uzbekistan; 13 officials representing the local organizers and the Scientific Information Centre of ICWC; and 12 international experts representing six international organizations working in the subregion, namely Canada International Development Agency, Swiss Development Cooperation (SDC), United States Agency for International Development (USAID), the World Bank, the Asian Development Bank (ADB) and IWMI.

d. The Subregional Workshop for South Asia on Application of Guidelines on Strategic Planning and Management of Water Resources in cooperation with IWMI and the Interim

National Water Resources Authority (INWRA) of Sri Lanka in Colombo from 10 to 13 September 2002. The subregional workshop was attended by totally 25 participants, including 20 officials from the eight countries of South Asia, namely Bangladesh, Bhutan, India, Islamic Republic of Iran, Maldives, Nepal, Pakistan and Sri Lanka; two participants representing the local NGOs, one international expert provided by the Department of Foreign International Development of the United Kingdom and two officials of IWMI.

e. The subregional workshop for North-East Asia was organized in cooperation with the Ministry of Water Resources of China in Beijing from 22 to 25 October 2002, which consisted of two parts: the International Forum and Subregional Workshop for North-East Asia on Application of Guidelines on Strategic Planning and Management of Water Resources. The Symposium was attended by totally 55 participants, including 11 foreign participants and 44 senior Chinese officials representing various Departments of the Ministry of Water Resources and the four major river basin commissions. The international participants included senior officials from Mongolia, Russian Federation and invited experts from UNESCO, World Bank, IWMI, Stanford University, Australia and the United Kingdom. The workshop was attended by 14 officials, including the international participants. In all the subregional workshops, the participants discussed details of the Guidelines for water resources and various aspects of the application to the Guidelines to water resources management, including practicability of such application to the formulation of policies and strategies of water resources management as case studies at the country level.

c. Phase III of the application of strategic planning and management into water resources management

Based on the interest generated at the subregional workshops, subsequent consultations were made with the focal points of interested countries. Case studies were then arranged for seventeen countries during the period from 2003 to 2004 for the development of strategic plans as follows:

a. Central Asia: Case studies were made for the five countries, Kazakhstan, Kyrgyzstan, Tajikistan, Turkmenistan, and Uzbekistan, within the framework of the Inter-State Commission on Water Management Coordination (ICWC) and in cooperation with ICWC Scientific Information Centre (SIC) for downstream areas of Syrdarya and Amudarya including whole Ferghana Valley. In addition, the Guidelines were also adapted using the experiences of these case studies for subsequent training of water resources engineers in the subregion. In this connection, one seminar was held in November 2003 in Tashkent in cooperation with SIC, IWMI, SDC and the United States Department of State to map out detailed application, a workshop to discuss the draft strategic plans in April 2004 in Tashkent and the training seminar was organized by SIC in November 2004.

b. North-East Asia: Case studies were conducted for two river basins in China (Haihe and Huaihe) in cooperation with the Ministry of Water Resources of China and for overall water resources management in Mongolia in cooperation with the Ministry of Environment of Mongolia. The final workshops were held in October for these two countries.

c. Pacific: Case studies were carried out for the Nadi River basin in cooperation with the Department of Land and Water Resources Management of Fiji and for the Laloki River basin in cooperation with the Water Resources Management Branch of Department of Environment and Conservation of Papua New Guinea. The final workshops were held in Nadi in September 2004 for Fiji and in Port Moresby in October 2004 for Papua New Guinea to discuss the draft strategic plans for the respective river basins.

d. South Asia: Case studies were conducted in Pakistan and Sri Lanka in cooperation with the Pakistan Council for Research in Water Resources (PCRWR) and the Interim

National Water Resources Authority (INWRA) of Sri Lanka, respectively. The final workshops were held in Islamabad in February 2004 to discuss the draft strategic plan for PCRWR and in July 2004 to review the draft strategic plan of INWRA.

e. South-East Asia: Case studies were conducted for the following countries: the Lao People's Democratic Republic on strategic plan of the Water Resources Coordination Committee (WRCC) in cooperation with WRCC; Malaysia on strategic plan for the Selangat River basin in cooperation with the Malaysia Water Partnership and the Department of Irrigation and Drainage; Myanmar on strategic plan for the coordination of water resources sector: establishment of the national water resources committee in cooperation with the Irrigation Department; Philippines on strategic plan for the National Water Resources Board (NWRB) in cooperation with NWRB; Thailand on strategic plan for the water sector - Department of Water Resources in cooperation with the Water Resources Association of Thailand; and Viet Nam on strategic plan for the Vu Gia-Thu Bon River basin in cooperation with the Institute of Water Resources Planning of Viet Nam. Six workshops were organized to discuss the respective draft strategic plan in July 2003 in Putra Jaya, Malaysia; September 2003 in Bangkok, Thailand; in December 2003 in Vientiane, the Lao People's Democratic Republic; in July 2004 in Manila, Philippines; in August 2004 in Da Nang, Viet Nam; and in September 2004 in Yangon, Myanmar. Subsequently, the Ministry of Agriculture and Rural Development of Viet Nam decided to hold a workshop to establish the Vu Gia-Thu Bon River Basin Organization on the basis of the strong recommendations of the case study and the related provincial authorities.

Lessons from the Application of SPM to Water Resources

The key results and outcomes of the project "Capacity-building in Strategic Planning and Management of Natural Resources in

Asia and the Pacific" in the water sector can be summarized in the following three aspects: important achievements of the project, key obstacles in application of SPM, and important lessons learned and best practices.

The important achievements of the project can be described in three main aspects: tool development, interest generation, and practical application of the methodology.

Tool Development

The development of the Guidelines marked an important step in the application of strategic planning and management approach to water resource, since it was the first time that a specific set of guidelines were specifically developed for the water sector and derived mainly from the practical experiences of the region. This fact can be illustrated by the translation of the Guidelines into Chinese, Russian and Vietnamese by the respective technical agencies. In addition, several enquiries were addressed on the guidelines on application to water resources management in the forestry sector in Malaysia and to the development of integrated river basin management modules for the UNESCO-IHE Institute for Water Education.

Generated Interest in SPM

The interest generated by the project on SPM of water resources was higher than expected in terms of the number of requests for case studies, which was higher than the project could deliver, although the actual number of case studies was much higher than the original number of case studies planned for the project, 17 in comparison with 10. Of particular interest was the decision of the Vice Minister of Water Resources of Uzbekistan to immediately apply the concept at the subregional workshop.

Best Practices of SPM

The seventeen case studies on the application of SPM to the water resources sector were conducted for the practical purpose of formulation of related policies and strategies at all the levels as recommended at the regional and subregional workshops: national (Myanmar, Thailand, Sri Lanka, Mongolia), basin (China,

Fiji, Malaysia, Papua New Guinea, Viet Nam and five countries in Central Asia), sectoral level (Pakistan) and organizational (the Lao People's Democratic Republic and Philippines). Out of these case studies, the momentum generated by the application for the case studies in Fiji, Myanmar and Viet Nam proved the immediate impacts of the project, in terms of immediate actions taken by the agencies concerned.

PRINCIPLES ADOPTED FOR APPLICATION OF SPM TO WATER RESOURCES

Throughout the implementation of the project on "Capacity-building in Strategic Planning and Management of Natural Resources in Asia and the Pacific", it was recognized that water resources development planning is a continuing process and the introduction of SPM into the national development process would thus need to be integrated into the ongoing efforts to improve water resources development planning. In that context, the principles for application of SPM to water resources were drawn up on the basis of the importance or necessity of SPM, key elements or components of SPM, and initiation and implementation of SPM.

Necessity of SPM for Water Resources Development

Throughout the implementation of the project on "Capacity-building in Strategic Planning and Management of Natural Resources in Asia and the Pacific", all participating experts and institutions recognized the importance and necessity of the adoption of SPM approaches in water resources management to ensure effective integration of water resources management into the national development process, to achieve efficient allocation of water resources against competing uses and to achieve consensus building and sustainable results of development. It was further recognized the importance of SPM in promoting good governance in water resources management and in capacity-building and improvement of management to cope with the increasing scarcity of water resources in the subregion. The participating experts also identified different priorities for the application of SPM to water resources at the national, sectoral and organizational levels.

a. At the National Level

Priority 1: (1) to provide frameworks and directions for: overall development, water sector and organizations, (2) to facilitate coordination among: agencies, sectors and stakeholders, (3) to improve implementation: funding, accountability monitoring, resources mobilization and HRD.

Priority 2: to bridge the present and future: to identify problems, to overcome present weakness, to respond to uncertainties and to enhance flexibility. Other: to share water among sectors and regions, and to achieve the targets within timeframe.

b. At the Sectoral Level

Priority 1: (1) to take into consideration different institutions, stakeholders, agendas of the different sectors (e.g. forestry, environment, lands, mines, protected areas) and encourage their participation, (2) to achieve expected outcomes within anticipated timeframe, (3) to identify common objectives and shared vision.

Priority 2: (1) to enable rapid development competing from limited resources, (2) to provide criteria for allocation of limited budget, (3) to provide guidance to tangible expected outcomes, (4) to coordinate long-term/short-term activities.

Other: (1) to improve legal framework from the many laws, (2) to re-audit performance, (3) to conserve water resources for sustainable use, (4) to address different target groups with different ability to pay.

c. At the Organizational Level

Priority 1: (1) to establish effective action plan, (2) to identify clear goal, (3) to create partnership to achieve mission.

Priority 2: (1) to establish success indicators, (2) to strengthen internal environment.

Key Elements/components and Issues of SPM

Through the process of application of SPM approaches to water resources management, the participating experts identified the following elements/components and issues as priority at the national, sectoral and organizational levels of water resources management:

a. At the National Level

Priority 1: (1) Element: clear vision based on shared concerns, (2) Driving forces: additional food requirements, additional area for irrigation, flood protection, and water needs for agriculture and domestic and industrial, (3) Issues: policy reforms, (4) Elements of SPM should cover technical, social, financial and environmental aspects.

Priority 2: (1) Issues: institutional reforms and mechanisms, public participation and resolution/prevention of conflicts/ differences.

b. At the Sectoral Level

Priority 1: (1) cross-sectoral approach, (2) relations between subregional and national priorities: transboundary water resources.

Other: (1) issues/problems, (2) fair allocation - explicit water rights, (3) restoration of water quality, (4) rational satisfaction of users, (5) environmental action and maintain natural equilibrium, e.g. watershed management, (6) decision-makers: to involve political masters and non-technical decision-makers, (7) participation of civil society.

c. At the Organizational Level

Priority 1: (1) Options of SPM: key steps and its example (country case), (2) Various approaches on SPM (collaborative, accountability), (3) Shared vision.

Priority 2: (1) SPM compare "conventional", (2) Glossary of key terms (vision, goal, objective, target).

Other: (1) Related government structures (level of administration, sectors, nature of water, etc.), (2) SWOT, (3) Environment, sanitary, land issues. In addition, the participating experts also recommended to address the following key issues in the process of application of SPM: politics, commitment, consensus building, and cultural diversity.

Initiation and Implementation of SPM

The experiences of the application of SPM approaches to water resources management of the project on "Capacity-building in

Strategic Planning and Management of Natural Resources in Asia and the Pacific" pointed out the importance of initiation of SPM processes at the national, sectoral and organizational levels as well as the need to maintain the momentum generated by the initiation. The participating experts of the project recommended the following key elements to be carefully examined as part of the principles to be adopted for the application of SPM to water resources.

a. At the National Level

Priority 1: (1) Political and decision-makers: willingness, commitment and concerns, (2) Implementation mechanisms: coordination, integrated into budgeting plans, strong auditing, monitoring and evaluation and team assignment, (3) Capacity building: training, HRD and experiences build up.

Priority 2: Legal basis for SPM: set up, enforcement and consultations.

b. At the Sectoral Level

(i) Triggering of SPM

Priority 1: (1) to base on initiatives of political and other groups at important benchmark events such as the World Water Forum, Agenda 21, etc. (2) to provide training and awareness programmes.

Priority 2: to promote political will.

Other: (1) to identify and give authority to a central coordinating body, (2) to identify champions, (3) to integrate SPM into policy directives.

(ii) Maintaining momentum of SPM

Priority 1: (1) to provide accountability and incentives, (2) to ensure cyclical/periodic review/reporting.

Priority 2: (1) to carry out the SPM process at national level, (2) to build consensus among knowledge base, (3) to implement identified tasks, (4) to include local perception.

(iii) Supporting measures

Priority 1: (1) to publicize successful case studies, (2) to be sufficiently flexible to incorporate necessary changes.

Priority 2: (1) to include financial auditing, (2) to provide legal support.

c. At the Organizational Level

Priority 1: (1) to provide "incentives" for the decision-maker of key actors of water stakeholders (such as political promotion, fund/budget), (2) to monitor performance indicators and accessible information, (3) to enhance public demand and appreciation.

Priority 2: (1) to promote educational awareness, (2) to provide incentive/promotion.

Other: (1) to identify champions, (2) to initiate at political platform, (3) to discipline offending, (4) to narrow the knowledge gap and enhance understanding of the future vision of the SPM process, (5) to pay special attention to great difficulties affecting a paradigm shift from traditional approach, (6) to persuade by using a clear vision and goals.

CONCLUDING REMARKS ON APPLICATION OF SPM TO WATER RESOURCES

At the Concluding Workshop of the project on "Capacity-building in Strategic Planning and Management of Natural Resources in Asia and the Pacific", held in Bangkok in November 2004, the participating experts recommended the following points to be considered in the application of SPM approaches to water resources management.

Contribution of SPM Approaches to the Development of IWRM Plans

SPM approaches were recognized as instrumental to the application of the concept of IWRM to support the socio-economic development process. While IWRM is acknowledged as a process, SPM was recognized as a powerful tool to enable priority activities of water resources management to be effectively integrated into the socio-economic development process through the following aspects:

- SPM enable priority actions to be strategically implemented.

- SPM is a process based on the scenarios analyses for the future, allowing uncertainties of the future to be incorporated into the planning process.
- SPM helps to regularly assess persisting issues.
- SPM promotes accountability, especially the core groups mainly responsible for the implementation of SPM.

It was considered that application of SPM to IWRM is the only choice to meet the urgent needs of socio-economic development when countries face serious water issues. It was also recognized that the application of SPM to development of IWRM plans would facilitate the mobilization of political will and commitment of key stakeholders, promote resolution of conflicts and ownership, and enable development of practicable plans of action.

The participating experts also drew up the following experiences of their application:

- SPM can help train new generations of water professional who would understand the challenges multidisciplinary approach needed for IWRM.
- SPM helps dissect major issues into manageable issues/ matters.
- SPM develops capacity to understand the pressing issues in terms of socio-economic aspects and other frameworks.
- SPM helps establish a core group comprising members from different agencies.
- SPM promotes pilot river basins.
- SPM is the effective tools to attract the assistance from donors and shows clear areas of responsibility, where stakeholders are participating.
- SPM can provide catalytic elements for change management and transform transitional institutions to think in terms of real time situation.
- Political will/government support and financial resources are needed for a successful SPM.
- SPM helps in raising awareness among various stakeholders.

Expectations from SPM

The participating experts in the Concluding Workshop mentioned above also identified the following expectations from the application of SPM approaches to water resources management as a possible checklist:

a. Change in Mindsets

Successful application of SPM approaches to water resources management is expected to result in the change of mindsets in the following areas:

- Development concept: socio-economic
- Water concept: IWRM, water utilization, ecological rehabilitation, flood mitigation
- Implementation: mobilization of support and commitment of policy decision-makers
- Foundation of water management: improvement in the understanding of the "reality" of water management systems as the basis for integration to form the basis to see how things actually work, i.e. going beyond institutional and other frameworks
- Efficiency of water resources management: change in mindsets to conserve water resources

b. Mechanism of Work

- Use SPM to develop network: SPM can be used as a tool generate consensus building, such as possible guidelines to develop action plans for food security and poverty alleviation
- Strategic Plan of Management of Water Resources: must take into cognizance Water Resources assessment within a river basin. In many countries, economic development is based on administrative land. What are areas of collaborative planning that can be developed, particularly economic, among stakeholders
- Increase networking through partners and among stakeholders

- Better decision-making in the process of water resources management
- Task force for collaboration is also essential
- Integration of IWRM to national social and economic development strategy and action plan
- To set up new legal institutions, needed for better management
- Building up of an effective coordinated block of all water related organization
- Improved water resources management structure of the Government

c. Implementable Programme of Action

- Ownership and sense of belongings of plan of actions developed by each basin
- Coordinated manner between social, economic and environmental aspects
- SPM should help us in identifying priority actions also at sectoral level
- The devil is in the details; having good SPM should help us go into the details of what to do
- Collaborative planning to achieve IQRM focusing on elemination of side effects due to sectoral focus
- To use the limited resources (water, human, money, etc.) in the best way
- Identification of areas need to be developed/improved

d. Mobilization of commitment for support and participation

- Instill in every stakeholder's mind/citizen on future scenario based on vision developed by them
- Enhance organizational effectiveness of the water resources coordination system and promote capacity-building
- Set up and develop the water market
- Civil society involvement in water resources management

EXAMPLES OF GOOD PRACTICES

The gross domestic product (GDP) of China increased substantially in the early 1990s, with industry and manufacturing contributing a large proportion of this growth. The rate of GDP growth was recorded at 6.5 per cent from 1960 to 1970, 9.7 per cent between 1977 and 1985, and 8.7 per cent in 1996 and 1997. The growth rate in foreign trade was equally impressive during the same periods at 14.3 per cent, 15.2 per cent and 7.6 per cent, respectively.

In the Five-Year Plan (2000-2005), the Government of China has set the objectives of developing and maintaining a high quality of life for all. In the process of developing the management of its strategic plan, the Government has specified the following five goals:

- Achieving an average economic growth rate of 7 per cent annually, which will translate into a per capita GDP of RMB 9,400 by 2005;
- Optimizing the industrial infrastructure, and strengthening international competitive capacity;
- Increasing research and development expenditure to more than 1.5 per cent of GDP, and improving education;
- Controlling population growth to keep it within 1.33 billion by 2005, improving the eco-environment and conserving resources; and
- Achieving an average annual per capita income of about 5 per cent, and improving health and social welfare.

These development goals clearly require support from the water sector, which comprises a number of agencies with resource development and management functions related to water resources. The sector is largely structured around the various uses of water, such as hydropower, irrigation, water supply, fisheries and wastewater. These water subsectors have developed their programmes with the aim of providing the support required for China's development objectives. These programmes, however, do not consider the potential for interaction between the policies and plans of the subsectors, because no agency was given overall responsibility for water resources management or coordination of water development.

China Water Sector

China is located in south-eastern Eurasia and covers an area of 9.6 million km^2. Affected by ocean-land distribution and topographic conditions, China is characterized by a monsoon climate. South-eastern China is pluvial while the north-western region is dry with the main period of rainfall being concentrated in the summer season.

China has a large number of rivers, totalling some 50,000 with catchments larger than 100 km^2 and 1,500 rivers with catchments larger than 1,000 km^2. Most of the rivers are located in south-eastern China. In addition, the country has about 2,300 lakes larger than 1 km^2 in area, with a total water area and volume of some 71,790 km^2 and 709 billion m^3, respectively. The average annual precipitation amounts to about 6,200 billion m^3, equal to 648 mm, which is 20 per cent less than the world average. Runoff amounts to 44 per cent of total precipitation; that is, China has an annual runoff of 2,712 billion m^3, equal to 284 mm. In a moderately dry year (1 in 4 recurrence) runoff falls to 2,549 billion m^3, while in an extreme drought year (1 in 20 recurrence) runoff amounts to only 2,359 billion m^3. The total average groundwater resources are estimated at 829 billion m^3, but the estimated overlap between runoff and groundwater is 718 billion m^3, giving an average total volume 2,820 billion m^3 of water resources. Total annual water resources amount to 2,840 billion m^3, accounting for one-sixth of the global total; however, per unit area is 83 per cent of the world average. The per capita amount of water resources is 2,210 m^3, equivalent to one third of the world average. Water resources per mu (cultivated land) amount to 1,440 m^3, equivalent to about half of the world average. Water resources are, in fact, scarce in China.

Although the total amount of water is more than adequate, distribution is uneven, in terms of both location and occurrance. Water distribution does not match land resources and fertility allocation. In southern China, total area is 36 per cent of the country, population is 54 per cent, cultivated land is 40 per cent and GDP is 56 per cent, while the water area is 81 per cent of the total. In northern China, the area is 64 per cent, population is 46 per cent, cultivated land is 60 per cent, GDP is 44 per cent, but water amounts to only 19 per cent of the total.

In the Yellow, Huaihe and Haihe River basins, the total area is 15 per cent, cultivated land is 35 per cent, population is 35 per cent, GDP is 32 per cent, but water is just 7 per cent of the total, and the per capita water resources is only 457 m3. In this region, there is the most critical conflict between water supply and demand. The monthly distribution of the flow of the rivers closely follows the pattern of rainfall: about 60 to 80 per cent during the wet season, especially in northern China, where rainfall is more concentrated.

The two processes of natural water inflow and water demand do not occur in harmony. Water is an important factor affecting balanced development among the different regions, populations, resources and environments in China.

Despite the uneven distribution of water resources, they are able to support economic and social development in China through the optimization of water resources allocation, both in terms of location and occurrance.

Scope of the Study

Strategic planning and management (SPM) aims to maximize the benefits to China through its water resources planning and management function. This study considers the component plans of SPM in China as specified in the draft policy, including:

- The principles of water resources management
- Water resources development and management
- Public involvement
- Financial support for water resources development and management
- Water allocation and quality management
- Data and information management
- Capacity-building and human resources improvement

SPM is carried out on the basis of a careful review of past development. In particular, it considers the issues related to the Haihe and Huaihe River basins, including flooding and water logging, water shortage, ecological deterioration and water pollution. The formulation of the development and management strategic plans follows a consultative process in order to obtain the views of the stakeholders as well as the Government of China and its agencies.

Water Resources Potential and Challenges in the Haihe River Basin

Although the Haihe River basin is facing a water resources shortage, it has much potential for supporting socio-economic development by efficient water usage and allocation (e.g., diversion), particularly in the irrigation and water supply subsectors.

So far, 32 large-scale reservoirs and 190 medium-sized reservoirs have been constructed within the Haihe River basin, covering more than 85 per cent of this mountainous basin area and providing a total reservoir volume of 29.4 billion m^3. More than 18,000 water diversion and pumping projects have been completed. The large projects include the Beijing-Miyun Water Diversion Project as well as the Water Diversion Projects from (a) the Luanhe River to Tianjin (and Tangshan), (b) the Qinglong River to Qinhuangdao, (c) the Yellow River to Hubei Province, (d) the Yellow River to the northern part of Shandong Province and (e) the Yellow River to the northern part of Henan Province.

A total of 1,200,000 motor-pumped wells have also been constructed.

In 1998, the actual gross water supply within the whole basin was 43.2 billion m^3, including 26.5 billion m^3 of groundwater (rated at 61 per cent), 11.2 billion m3 of local surface water (26 per cent), 5.1 billion m^3 of Yellow River diversion water (12 per cent) and 400 million m^3 from other sources (brackish water, seawater and treated wastewater) (1 per cent). When classified according to the different water consumption sectors, 2.6 billion m^3 was for urban domestic (6 per cent), 140 million m^3

for improving city river and lake environments (0.4 per cent), 7.05 billion m^3 for industry (16.3 per cent) and 33.41 billion m^3 for agriculture (including irrigation, rural living, forestry, livestock and fisheries) (77.3 per cent).

However, water resources exploitation is at a very high level: groundwater is over-extracted, downstream reaches are dry, and water diversion is needed from outside the basin. In addition, several other problems face the water sector, including unusual rainfall patterns in some years, high evaporation rates, flooding and drought in the basin.

Being an over-developed river basin with limited water resources, the most important challenges facing the Haihe River Basin Water Resources Commission (HRWRC) in carrying out its coordinating role include:

(a) The need to strengthen the legal framework for effective and harmonious integration of water resources management, development and protection activities into the socio-economic development process of the basin, especially to meet national priorities;

(b) The need to enhance and consolidate the existing systems as well as provide a foundation for operating, maintaining and rehabilitating facilities safely, reliably and efficiently in order to protect the investment for public benefit; and

(c) Prioritize the capacity-building needs in order to enhance the organizational capacity and effectiveness of the water resources coordination system.

Water Resources Potential and Challenges in the Huaihe River Basin

The Huaihe River basin is situated in eastern China at longitude 112 -121 east and latitude 31 -36 north, between the Yangtze River and the Yellow River with a total area of 270,000 km^2. The runoff of surface water changes greatly from year to year. The mean quantity of surface water resources is 66.1 billion m^3, while groundwater totals 40.6 billion m^3 and net water resources amount to 96.1 billion m^3. In 2000, withdrawals of water resources totalled 61.2 billion m^3, with exploitation reaching 63.7 per cent.

The water in the Huaihe basin is heavily polluted. The stretches of the river with a water quality worse than class IV is 69 per cent of the total river length, causing water scarcity in the affected areas. The Huaihe basin is also threatened by severe flooding during the monsoon season.

Summary of Goals Linked to the Socio-economic Development Goals

a. Overall Socio-economic Development Context of China and the Haihe River Basin

China is currently undergoing a dual transition from a rural to urban/industrial society and from a command to a market economy in order to integrate into the global economy and promote economic growth. Forces promoting this transition are political and macroeconomic reforms implemented through policy and legal decrees in the areas of (a) state-owned enterprises (SOEs), (b) finance, (c) decentralization, (d) labour markets, (e) government and (f) trades.

The implementation of these reforms is affecting social trends including (i) changes in population growth, (ii) urbanization, (iii) rising income and income disparities, and (iv) rising consumption of goods and services. Those social trends and economic reforms combined with natural and historical conditions have various impacts on water resources and the institutions that manage them. Complex technical, social and management solutions are required for addressing the growing water shortages and pollution.

According to the tenth Five-Year Plan (2001-2005), the economic growth rate is expected to reach 7 per cent; that is, GDP will be RMB 12,500 billion in 2005, in which the shares of agriculture, industry and service sectors will be 13 per cent, 51 per cent and 36 per cent, respectively. The population growth rate is expected to slow to less than 0.9 per cent in 2005 with a total population of some 1.33 billion.

The population of the Haihe River basin is 118 million, representing 10 per cent of the national total, of which the urban and rural populations account for 24 per cent and 76 per cent, respectively. The average density is 371 persons/km2, which is 3.47 times the average national density. Moreover, there are large differences in density between provinces (cities) in the basin, ranging from 56 persons/km2 to 784 persons/km2. At present, there are two municipalities directly under the central Government, five provinces and an autonomous region, including 65 regional administrative districts (Beijing and Tianjin comprise 36 districts and counties), all of which are at the regional level and 235 administrative districts at the county level.

Cultivated land in the Haihe River basin totals 163 million mu, which is 11 per cent of the national total. Of this area, dry land accounts for 158 million mu and irrigated fields for 4.79 million mu. The effective irrigation area is 102 million mu. Cultivated land per capita is 1.38 mu, equal to the national average value. From 1949 to 1993, cultivated land decreased by 23.2422 million mu, hence 14 per cent of present cultivated land, or an average annual decrease of 520,000 mu, hence 0.3 per cent of present cultivated lands. Therefore, cultivated land should be valued. Moreover, the remaining undeveloped 40 million mu can be developed.

In Haihe, the total value of agricultural output is RMB 97.5 billion, while total food yield is 45.4 million metric tons, accounting for 10 per cent of the national total. Food per capita is 385 kg. The gross value of industrial output is RMB 644 billion (of which RMB 270 billion is from village/town enterprises that account for 36.4 per cent), representing 15 per cent of the national total. The total value of industrial and agricultural output combined is RMB 742 billion, while overall per capita income is RMB 6,285.

The Haihe River basin is rich in mineral resources such as coal, petroleum, gas, steel, aluminium, gypsum, graphite and sea salt. For example, coal reserves are some 202.6 billion mt, which is nearly 30 per cent of the national coal reserves; annual extraction is 28 million mt, or 20 per cent of the national figure. Moreover, the Huabei, Dagang and Shengli oilfields as well as part of the Zhongyuan oilfield are in the Haihe River basin, with 1.5 billion mt of oil reserves and 36 million mt of annual output.

The Jing-Guang, Jing-Jin, Jing-Ha, Jing-Hu, Jing-Bao and Jing-Yuan railway lines cut through the Haihe River basin. According to available but incomplete statistics, in Beijing City, the volume of passenger traffic is 110 million persons per year, while the volume of freight traffic is 460 million mt. Along the Bohai Sea, there are the ports of Qinhuangdao, Tanggu and Huanghua, among others, with 109 berths and a handling capacity of 140 million mt. The air transport centre is Beijing, with airports in Tianjin, Shijiazhuang, Qinhuangdao, and Xingtai and Changzhi, accounting for 4.16 million passengers and 2.98 million mt of freight traffic annually.

The vast Huabei Plain, with its good climate and abundant sunshine, is one of the main grain producing regions. The food crops comprise mainly wheat and corn, plus Chinese sorghum, rice, millet and beans. Industrial crops are cotton and oil plants. In 1993, the total value of agricultural output was RMB 94.47 billion, and the average grain yield per mu was 292 kg. Currently, productive power is weak with the exception of a high output in the alluvial plain and Yellow River diversion irrigation region. The Haihe River region is also rich in fruit, and its forestry productivity and stock-raising can be developed further, combined with water and soil conservation. Moreover, there are 3 million mu of saline-alkaline wasteland and many reservoirs, and the low-lying lands should be developed.

b. Opportunities and Threats for Water Resources Development from the Perspectives of the Haihe River Basin and Related Provinces

With more than 18,000 water diversion and pumping projects having been completed, including large projects such as the Beijing-Miyun Water Diversion Project, the Water Diversion Project from Luanhe River to Tianjin municipality and Tangshan City, and the Water Diversion Project from the Yellow River to the northern part of Henan Province.

In 2000, the actual gross water supply for the basin was 40 billion m3, including 26.3 billion m^3 of groundwater (about 65.8 per cent), 13.6 billion m^3 of surface water (including diversions, accounting for 34 per cent) and 100 million m^3 of water supplied from other sources (brackish water, seawater and treated sewage). Overall water consumption by sector is 5.2 billion m^3 for domestic consumption (some 13 per cent), 6.6 billion m^3 for industrial uses (some 16.5 per cent) and 28.1 billion m^3 for agricultural uses, accounting for 70.3 per cent of the total.

With this demand/supply balance, the annual water flow rate to the sea decreased from 24 billion m^3 in the 1950s to 6.7 billion m^3 in the 1990s. The reduced flow has caused serious sedimentation in the river mouths and estuaries.

For the past 10 years, basic water demand for socio-economic development could only be satisfied by over-extracting by some 5-7 billion m^3 of groundwater and re-using nearly 3 billion m^3 of untreated sewage water. The shallow groundwater over-extraction area is presently about 44,000 km^2; however, in some 10,000 km^2 of the sediment area the shallow groundwater resources have dried up.

The deep groundwater over-extraction area is 56,000 km^2. The total over-extraction area of both shallow and deep groundwater resources in the whole basin is estimated to be 89,000 km^2, accounting for 69 per cent of the plain area of the basin. Consequently, environmental and geological problems, such as land subsidence, have been induced in the areas of extreme over-extraction.

Due to the excessive utilization of surface water resources, many rivers dry up in most seasons or are now dry throughout the year. In addition, over-discharging of pollutants into the rivers has left much of the remaining water seriously polluted. An example from 1998 best illustrates the serious water situation. During that year, total precipitation was 175.3 billion m^3 (equivalent to 551 mm in depth) and the gross volume of water resources amounted to 35.5 billion m^3, which represents an average year.

However, the water resources utilization ratio was 90 per cent, and the total water flow to the sea (including untreated sewage) was 4.3 billion m^3.

c. Vision Statement of the Haihe River Basin Water Resources Commission

By consulting representatives from HRWRC, the Ministry of Water Resources and some of the provinces concerned, the following vision statement for HRWRC has been determined: "Towards a most modern, reliable and efficient water resources management system to protect, manage and develop the World's largest complex water transfer system to achieve sustainable well-off life in the Haihe River basin."

The highest organ of the Government is the National People's Congress (NPC). Of nine special committees established by NPC, two are of particular importance to the water sector: the Environment and Resources Protection Committee, and the Agricultural and Rural Affairs Committee. The Chinese People's Political Consultative Conference (CPPCC) is the highest advisory body in the country, meeting on an annual basis and consisting of more than 2,200 representatives from all social sectors. Its primary function is political consultation, democratic supervision and participation in, and deliberating on State affairs.

Executive power is exercised by the State Council. Under the State Council are the functional organizations such as ministries and commissions. According to the official government structure, provincial level departments and bureaus of water conservation, and community level water conservation stations and water users associations are vertically integrated with the Ministry of Water Resources. However, they also report and depend on the provincial governments especially in relation to administration of laws and importantly for funding. This form of decentralization may be defined as "delegation" where the central government transfers responsibilities for decision-making and administration of public functions to local governments.

The "principled laws" that guide the activities of the ministries in water resources management or environmental protection are based on well-established scientific doctrines and China's own blend of water administration law, enforcement and jurisdiction. These well-meaning principles (such as water allocation based on watershed or polluter pays principles) increasingly conflict with the economic interests of the provinces. Those administrative bodies (whose financial viability depends on provincial budgets) empowered by law to enforce rules and regulations at the local level (such as water bureaus) are often under pressure to act in the interests of the local government to the detriment of sound water or environmental resource management and planning. Thus, vertical relationships between ministries, departments, bureaus etc. have been challenged by stronger alliances between the local governments and these local administrative bodies.

In an effort to improve holistic planning for water resources, coordination among ministries, commissions and lower level government entities is effected through "leading groups" (coordination committees) or interdepartmental agreements. The main functions of the leading groups are to:

(a) Examine and approve the comprehensive planning of large river basins

(b) Examine and approve the important principles and policies of national water and soil conservation work as well as the important problems of key control projects

(c) Resolve major water resource allocation problems among various sectors

(d) Handle and coordinate major water affair conflicts between provinces

The only standing inter-ministerial committee in the water sector at the national level is the State Flood and Drought Control Relief Headquarters. During times of emergency, this office reports directly to the responsible vice-premier. A temporary coordinating committee may be established for a particular purpose (e.g., related to a specific basin or programme or issue), and if so, is chaired by the appropriate ministry. Provinces may also establish leading groups for specific programmes or issues within their jurisdiction, although again this practice may now be discouraged.

As at the central level, provincial planning commissions are responsible for reviewing and approving programmes and projects, and may play an important role in resolving wider coordination difficulties. In practice, there is considerable overlap between the different ministries, and inconsistent and fragmented responsibilities are a major issue. While legislation calls for integrated and comprehensive management, in practice, different ministries and agencies are primarily responsible for implementing specific acts of legislation; thus, inconsistencies and conflicts arise between different agencies in the exercising of their respective mandates. This is probably why resource conflict mediators and interprovincial water allocation planners such as leading groups were set up in the first place. However, the political process of decentralization (which has undoubtedly helped to boost China's economic growth) has taken away this high-level planning perhaps because it is viewed as being too centralized.

In addition, with the intensification of competition between provinces due to (a) increasing economic activity, (b) population growth and (c) new imperatives of regional responsibility systems, local governments have had to acquire planning/management rights to natural resources in order to sustain economic growth. Micro planning and management at the regional level, limited by administrative boundaries, is thus the current reality in the water sector. Certainly, ministries have retained significant input into planning and management, but as central government revenue and public expenditures have shrunk, but they are constantly constrained by the limited resources available for public investment in national priorities and nationwide externalities such as large water transfers, wastewater treatment plants etc. These key projects may have a far greater significant impact on growth across provinces than their counterparts in each province. In addition, given the nature of water resources as a unitary good, multiple planning and regional management may have negative effects on economic growth.

Apart from inter-ministerial leading groups, other mechanisms designed to improve water resources management include the river basin management groups.

The Water Law recognizes the basin as the logical context for devising solutions to water resources management problems. How best to coordinate activities based on hydrological boundaries with those based on administrative boundaries is a perennial issue in water resources management. Among the issues that have arisen are the following:

(a) The River Basin Commissions (RBCMs) are commissions in name only, having no separate governing board or corporate status. RBCMs are departments of the Ministry of Water Resources and perform those functions that the ministry delegates to them. RBCMs find it difficult to enforce provisions of basin plans on other sector ministries and provincial governments, and the functions that they perform overlap with activities undertaken at the provincial and local levels;

(b) RBCMs in principle help resolve conflicts between jurisdictions and sectors, and ensure that multiple uses are served according to established priorities. However, these functions are hampered by an absence of formal agreements on interprovincial water allocation, pollution limits, and other matters. Only in the case of the Yellow River has water been allocated among the provinces and, even here, provisions for varying river conditions are inadequate. RBCMs may propose and (to the extent that they undertake the real-time management of multipurpose facilities) enforce allocations; however, most storage and diversion facilities are controlled by local or sector entities, and these may be operated in ways that are inconsistent with the provisions of RBCM plans. Only during times of flooding do RBCMs exercise predominant authority over other entities under the State Flood Control and Drought Relief Headquarters;

(c) The administration of land, groundwater and water quality is typically handled by agencies at the provincial and local levels. The transfer of aspects of groundwater administration to the Ministry of Water Resources and the provincial water resource bureaus may facilitate integration of groundwater and surface water administration, but conjunctive management as well as the integration of land

and water, and water quantity and quality, all remain illusive within the basin context. Little effort is made to consolidate plans covering each aspect within comprehensive county, provincial or basin plans; and

(d) In principle, RBCMs prepare basin development and operating plans, and undertake other tasks in full consultation with the provinces, sectoral ministries and other stakeholders. In practice, there are few formal consultation mechanisms, and the main directives and decisions affecting RBCM activities are received vertically from the Ministry of Water Resources.

Legal and Institutional Framework

a. Legal and Institutional Context

In 1954, the first Constitution was promulgated and was the foundation document for the socio-economic management of the country. It provided the basis for a multi-sector economy as well as the conditions for the education, culture and health of the Chinese people. Thereafter, the second Constitution was enacted in 1975, the third in 1978 and the fourth in 1982. Since 1982, the Constitution has been revised four times in accordance with the changing domestic, regional and global environment. In March 2004, the Constitution was revised in order to meet the real needs of the Chinese people in the current economic environment.

The first Water Law was formulated by the State Council and adopted by the National People's Congress Standing Committee in 1988. It was revised in 2002 as both the water resources conditions and water utilization situation had changed considerably. The Water Law sets out a legal framework for development of the water sector. The development of further legislation or a decree on subsector activities is urgently needed as well as the necessary legal documents to make the law effective. External assistance is also required by water subsector agencies in the drafting of regulations covering the operation and enforcement. A prerequisite for success in this area is capacity-building throughout the water sector, which traditionally focuses on development rather than management and regulation.

In addition, other laws related to water include the Flood Control Law, Water Pollution Protection Act, Water and Soil Conservation Act, Environment Protection Act, Land Management Act. The Flood Control Law was enacted to (a) prevent and control flooding, (b) alleviate calamities caused by flooding and water logging, (c) maintain the safety of people's lives and property, and (d) safeguard the smooth progress of construction modernization.

The Water and Soil Conservation Act was formulated for the purposes of (a) preventing and controlling soil erosion, (b) the protection and rational utilization of water and soil resources, (c) the mitigation of flood, drought and sandstorm disasters, (d) the improvement of the ecological environment and (e) the development of production. The Water Pollution Protection Act was enacted in 1996 to (a) control water pollution and protect the environment, (b) guarantee the health of the people and (c) promote the modernization of construction.

The Water Law grants the State Government the right to apply water licences and fees for water use. The Water Administrative Department of the State Council is in charge of the organization and implementation of water licensing and fees.

The State Government has adopted an approach that combines river basin management with district management. The Water Administrative Department of the State Council, which is supported by national laws and administrative decrees, is responsible for the integration of management and supervision of national water resources including important rivers and lakes defined by the State. It is also in charge of the exploitation, utility, efficient use and protection of water resources. The Ministry of Water Resources is responsible for integrated management and supervision of national water resources.

The provincial, municipal and county level governments are in charge of the integrated management and supervision of water resources within their districts, according to the Water Administrative Department.

The relevant departments of the State Council are responsible for the exploitation, utility, efficient use and protection of water resources. The provincial, municipal and county level governments are in charge of the exploitation, utility, efficient use and protection of water resources within their districts.

HRWRC, as an administrative agency of the Ministry of Water Resources, is authorized to undertake water resources management and supervision and is supported by national laws, administrative statues and the Ministry of Water Resources.

b. Current Situation and Perspectives of the Legal and Institutional Framework

The Ministry of Water Resources is a Water Administrative Department of the State Council. Other relevant departments include the State Environment Protection Administration, the Ministry of Land and Resources, the Ministry of Construction, the State Forestry Authority, the National Development and Reformation Committee, the Ministry of Communications, the Ministry of Public Health, the Ministry of Finance and the State Meteorology Authority. The Ministry of Water Resources grants its seven river basin administrative institutions the authority to manage the water resources within their regions.

As a subordinate body of the Ministry of Water Resources in the Haihe River basin and northern part of Shandong Province, HRWRC is empowered to administer the above-mentioned areas with regard to (a) managing the water resources and river course of the local basin, (b) control and develop important water conservancy projects, (c) undertake planning, management, coordination, supervision and advancement of river control, and (d) ensure all-around development, utilization and water resources protection.

HRWRC comprises a water administration and management department, an hydraulics and hydropower administrative department (flood prevention and drought prevention office), a department of water and soil conservation, a water resources protection bureau, an hydrology office, the Haihe River lower reaches administration bureau, the Zhang-Wei south channel administration bureau and the Zhanghe upper reaches administration bureau.

It should be noted that there is no clear river basin law or regulation that sets out standards for how a river basin administrative agency should carry out water resources management and supervision, despite the functions explicitly defined by the Water Law.

Development of clear legislation, regulations and guidelines is one of the primary means by which HRWRC can play its coordinating role within the water resources sector of the Haihe River basin. The process must be handled in an open and cooperative way, involving the stakeholders at the central, provincial and local levels. HRWRC has already gained some experience in this aspect, but further capacity-building for policy analysis and development is required. Legislation and institutions need to be supported by detailed implementation plans that indicate agency responsibilities and, where necessary, further capacity-building to allow implementation to be successfully carried out.

c. Strengths and Weaknesses of the Current Legal and Institutional Framework

Enforcement measures and detailed rules and regulations are lacking for the Water Law, Law of Water and Soil Preservation, Law of Environment Protection and Flood Control Law. In addition, some connection problems exist between local rules and regulations and the administrative management system. Some local rules and regulations are not in accordance with those of the State, and there are no special procedures for the inspection and coordination of enforcement measures at the local level. When enforcement measures are introduced, prior opinions are not sought from HRWRC, which makes it difficult for HRWRC to maintain control.

Although the water legislative position, legislative supervising authority and legal duties of HRWRC are defined in the existing Water Law, it is not specific. There are no effective management measures for ensuring the performance of the river basin plan, such as water resources development, allocation, supervising management, basin control, project approval, and engineering investment. Even if a project is in violation of the law, RBCMs are not granted the authority to take effective measures for preventing illegal activities, and the performance of the basin plan also fails to be effective. Thus, it is clear that the corresponding detailed rules and regulations for water resources management need to be formulated, such as those concerning water quantity and quality, and flood control. At present, the following problems require urgent attention:

- The fundamental nature, position and responsibilities of RBCMs, and authority to enforce the law.
- The relationship between the river basin management and administrative sectors.
- The management principles of the river basin and the basic management system.
- Policy measures for promoting the comprehensive development of the river basin.

Despite its title, the existing River Basin Council is only an agency of the competent water administrative department of the State Council. In fact, management is not carried out by the officials of the relative departments and sections. At present, following the establishment of the flood control headquarters and water resources protection agency, a great deal of successful experience has been gained.

Although the RBCM departments are mainly administrative institutions and deal with few enterprises, their institutional expenses are allocated by the financial department; however, the amount provided is less than the half of actual expenditures. Without sufficient financial support, an RBCM will be unable to handle its responsibility.

While an RBCM is in charge of some water supply and hydropower projects it has no authority to decide on the water and electricity prices. In effect, the charges for water and electricity are much lower than their costs; however, neither compensation for resultant losses nor starting and development funding are provided by the Government.

The management model that takes the river basin as a basic management unit for water resources will be unable to achieve efficient impact unless the management model used for water project funds provided by the State is the same as that used for a river basin. Together with the reformation of the State planning system, financial support by the central Government for such projects will be greatly reduced.

The current institutional problem in the water sector is related mainly to the lack of coordination between agencies within the sector and with those of other sectors, as well as the loose communication and coordination between the national agencies and their provincial counterparts.

The first step is to assess the present institutional set-up at the provincial and river basin levels, and to conduct a diagnostic study of each targeted river basin. The findings of these studies could be used as a basis for designing a river basin management body tailored to the development mix in each basin.

d. Haihe River Basin Water Resources Commission Mission Statement

The mission statement suggested by the water sector is "Protect, manage and develop water resources to achieve good living standards in the Haihe River basin".

As with the overall vision statement, the words have specific meanings. Protect refers to the need to ensure the continued supply of high-quality water resources. Manage refers to the water resources management activities, including access to water resources and the right to utilize water resources, and managing those institutions with responsibilities related to water resources.

Develop refers to converting water resources to new uses and to a mature condition for the purpose of improved quality of life.

The Circular of Functions and Duties of the Haihe River Water Resources Commission, No. [1994]24, published on 16 March 1994 by the Ministry of Water Resources, defines the rights and duties of HRWRC. Although broad, the list of rights and duties is mainly directed towards actions, such as "planning, managing, coordinating, supervising and serving" as follows:

- Supervise the enforcement of the Water Law and the Water and Soil Conservation Law, and formulated policies and regulations for the river basin;
- Work out medium- and long-term water resources development plans for the Haihe River basin, formulate a comprehensive plan and specialized plan together with the relevant departments and provincial government, and supervise enforcement of the plan after approval;
- Organize water resources monitoring and assessment in an integrated manner, formulate a long-term water supply and demand plan and a water distribution scheme, and undertake the supervision and management of water resources protection;
- Manage the river, lakes, river mouth and key river course in the basin, and examine construction projects;
- Formulate flood control plans for the river basin, supervise the flood control plans of trans-provincial rivers, coordinate the daily work of flood and drought control in the river basin, and supervise safety and construction in flood areas;
- Mediate water affairs among departments and provinces;
- Organize prevention, supervision and all-around control in areas where loss of water and erosion of the soil is serious;
- Examine the project suggestion report, feasibility plan and preliminary design of projects directly under the central authorities as well as local joint project;
- Be responsible for the comprehensive development of the river basin and the management of important trans-provincial water conservation projects;
- Direct the work of rural and urban water conservation, water conservation project and electrification project management; and
- Undertake other work as required and authorized the Ministry of Water Resources.

Strategic Plan Goals

a. Overall Haihe River Basin Water Resources Commission Goal

The overall HRWRC goal is to achieve consensus building and sustainable development of the water resources in the country, particularly in the priority areas identified by the Government. Within the context of this overall goal, an attempt will be made to achieve the following three integrated water and resources management (IWRM) main mission goals:

- IWRM Goal 1: To improve and modernize the legal and technical framework for managing, developing and protecting water and related resources in order to meet the needs of current and future generations;
- IWRM Goal 2: To enhance and consolidate the existing systems and foundation for operating, maintaining and rehabilitating facilities safely, reliably and efficiently in order to protect public investment; and
- IWRM Goal 3: To enhance the organizational capacity and effectiveness of the water resources coordination system.

The three mission goals are linked to the three priorities of the country mentioned at the beginning of the report, in terms of ensuring high economic growth, poverty alleviation and a strong framework for sustainable development.

b. Haihe River Basin Water Resources Commission Mission Goal 1

To achieve this mission goal, the Ministry of Water Resources will work in cooperation with all key ministries and stakeholders towards effectively managing, developing and protecting water and related resources for agricultural, municipal, industrial, rural, hydropower, recreational, fisheries and wildlife purposes.

The Ministry of Water Resources and the Haihe Basin Commission will pursue coordinating roles aimed at improving water and related resource management, development and protection activities. The framework for such coordinating roles would include:

(a) Legislation to improve integrated and comprehensive water resources management, including flood control, irrigation, urban and rural water supply, water pollution control and treatment;

(b) Legislation to strengthen the authority of the Haihe River Basin Commission to implement integrated and comprehensive water resources management. The Haihe River Basin Commission should be given more implementation authority to ensure sustainable water use, rather than just design and planning;

(c) Formulating comprehensive water resources plans, including surface water and groundwater, water quantity and quality, water use and protection, economic and environmental water use etc., which should get approval from the State Council and should act as guidelines of management activities;

(d) Formulating policy and strategies to improve the water resources planning systems of the country, with emphasis on river basin water resources planning systems;

(e) Formulating policy and strategies to improving the efficiency of water use; and

(f) Promoting integration of applied sciences and technology into water resources management.

c. Haihe River Basin Water Resources Commission Mission Goal 2

Under this mission goal, the Ministry of Water Resources and the Haihe Basin Commission will focus on operating, maintaining, and rehabilitating existing water resources facilities so that they continue to provide project benefits. These facilities provide power and water supply delivery systems that serve agricultural and municipal users, in addition to recreation, fisheries and wildlife benefits, and flood control.

Ensuring that the facilities are safe, cost-effective and reliable will assist the agencies concerned to operate them as effectively and efficiently as possible in order to provide project benefits while also protecting public health, sustaining environmental values, and providing timely and economical services to customers. This will protect the public investment.

To enable this goal to be achieved, the necessary legislation for ensuring safe and reliable water related services should be promulgated. At the same time, certain agencies should be endowed with the responsibility for maintaining the safety and reliability of such facilities.

d. Haihe River Basin Water Resources Commission Mission Goal 3

In order to achieve the above mission goals, HRWRC needs to be strengthened to make it more effective. The strategies for achieving the goal should include:

(a) The formulation of a reform plan for the Haihe Basin Commission to adapt to the increase in authority and the changed situation, with the focus on managing water resources in an effective and integrated manner. The reform plan will be presented to the Government for approval;

(b) The establishment of an information management system to facilitate management;

(c) The preparation of five-year Action Plans, based on the priority water resources management issues at the national level and river basin levels; and

(d) The preparation and implementation of a capacity-building plan for HRWRC, based on the requirements for achieving the Action Plan.

Implementation, Monitoring and Evaluation

a. Identification of Indicators and Benchmarks for Performance Measurements, and the Expected Timeframe and Necessary Resources

Implementation would be carried out based on the clarification of the strategic planning and management process.

(i) Human Resources Development

It is necessary to strengthen the River Basin Commission, set up a water resources coordinating committee, and develop the right skills and competencies to match the strategy, giving recognition to the fact that the Chinese people are the best source of ideas for their future. Moreover, confirmed leadership, creative teamwork and organizational culture are important to the successful implementation of strategic plan management.

Consequently, it will be necessary to:

(a) Build an organization around the critical strategic elements;

(b) Mobilize and allocate resources to match the strategic plan;

(c) Establish administrative systems, policies and procedures in accordance with the strategic plan;

(d) Develop a personnel reward system linked to achievement and the overall strategy; and

(e) Recognize that the most important job of leadership is to establish and instil confidence in the strategic plan management, and to ensure that everyone in HRWRC follows it.

Because of the complexity of strategic planning, it will be necessary to seek financial support, both at home and abroad. Only if HRWRC treats fund-raising as a fundamental part of the operation, and approaches it in a professional and systematic way, will strategic planning have any chance of becoming financially secure. As the Ministry of Water Resources has appointed HRWRC as the leading institution in the promotion and coordination of water resources development for the benefit of the river basin population, its duty is to treat fund raising not as an auxiliary issue, but as an essential component of the overall operation.

Financial security means that funding must be sought and obtained in a flexible manner from diverse sources at various levels, including the central Government and, in particular, the Ministry of Water Resources, local governments, communities and associations, involved enterprises as well as international organizations.

It is clear that a subsidy from the central Government will play a crucial role, while local governments would provide of counterpart funds. Financial assistance contributed by other institutions or organizations will be in a variety of forms, such as cooperative scientific research, transfers of project results at a preferential price or even free of charge. As for seeking overseas funding, this would take a similar course. In addition, HRWRC responsibility would include the identification of reasonable targets for funding and in-kind services during the process of strategic implementation as well as deciding on how the funds and services should be used.

With regard to financial management, HRWRC is accepted as the core authority in charge of all relevant issues. Its functions are clearly defined by the central Government, particularly the Ministry of Water Resources, and bolstered by local government at different levels.

The major duties performed by HRWRC are to (a) put forward the overall budget needed, (b) determine the framework of fund-raising and clarify the contribution percentages from different financial resources in the foreseeable future, and (c) establish and carry out efficient financial monitoring systems over the strategic planning period.

(ii) Action Plan

Three levels of action by HRWRC will be necessary, as described below.

(1) *First Level:* The first level includes the reasonable allocation of water resources, water conflict reduction and the establishment of a consolidated information system.

(a) Reasonable allocation of water resources. Water resources include surface water, groundwater, treated wastewater, brackish water and fresh seawater etc. Water use involves many aspects, such as domestic urban and rural usage as well as production and eco-environment usage. Optimized water allocation will be achieved via water allocation, coordination of water supplies and usage, unification of water quantity and quality, and real-time regulation and preference of the benefits. The mutual relationship of economic structures, economic development and water resources must be handled correctly via administrative, economic, technical and legal measures. Reasonable water resources allocation can ensure that limited water resources produce full benefits.

(b) Water conflict reduction. The River Basin Commission should coordinate the relationship between the provinces concerned, and between the different water-use sectors. By establishing an information sharing system, such a relationship can be formed based on mutual respect and trust. Using market mechanisms and establishing a compensatory mechanism are also important ways of reducing water conflicts.

(c) Establishment of a consolidated information system. On the one hand, the province will provide basic hydrological and water use information to the River Basin Commission. On the other hand, the River Basin Commission should provide information to the province and the public. This system will not only reduce water conflicts but also improve water use efficiency.

(2) *Second Level:* As detailed below, the second level of action includes water conservation and improved water-use efficiency, the use of seawater in coastal areas, the construction of a south-to-north water diversion project and the protection of groundwater resources.

(a) Water conservation and improved water use efficiency. At present, the industrial water saving level in Haihe River basin is very high, and the average water consumption quantum for every RMB 10,000 output value has been reduced from 160 m^3 in the early 1990s to 51 m^3. The water reuse rate is about 80 per cent, which is 40 per cent higher than the average national level. The efficiency of industrial water consumption can only be improved by adjusting the industrial structure and arrangements, limiting the high level of water consumption in production, adopting water-saving technology and gradually increasing the ratio of new technology industries.

Use of irrigation water in the basin is about 30 billion m^3, which is 70 per cent of the total water supply. The comprehensive irrigation quantum is some 296 m^3/mu. The water-saving irrigation area is 40 per cent of the total irrigation area, and it is planned to increase the percentage to 70 per cent in 2010. Together with the process of urbanization, domestic urban water use will be steadily increased. Domestic urban water saving will be realized mainly by popularizing water-saving appliances, raising the water price, reducing seepage losses in the running water conveyance pipe network, developing repeated usage and enhancing public awareness of water saving.

(b) Use of seawater in coastal areas. Haihe River basin has an approximate 900-km coastline. Therefore, the use of seawater is an important way of mitigating fresh water shortages in coastal cities in the future. At present, seawater is only used for cooling by the electric and chemical industries in Tianjin, Qinhuangdao and the coastal areas of Shandong Province. The use of freshened seawater lags far behind. Together with the continuous progress in seawater usage technology and the reduction of freshening costs, the use of seawater will have a bright future. Freshened seawater will eventually undergo even wider production.

(c) Construction of a South-to-North Water Diversion Project. The project being established is of strategic fundamental infrastructure for solving the water shortage problem in Haihe River basin. The two largest cities, Beijing and Tianjin, and the northern China plain are intake areas of the East Route and the Middle Route of the South-to-North Water Diversion Project. It is related to the domestic water supply as well as economic development and improvement of ecological environments in those areas. An analysis of the present and future water resources demand and supply in the basin shows that merely developing efficient water use and the potentialities cannot thoroughly change the water shortage situation and the worsening trends in the ecological environments. The water diversion project is necessary to meet water demand.

(d) Protection of groundwater resources. The groundwater resources of China, especially in the Haihe River basin, play a vital role in water resources that have served China well. The Haihe River basin has large areas where both shallow and deep aquifers are greatly overexploited. Approximately 28 per cent of the total groundwater used is from deep confined aquifers where the recharge is very limited. The heavy overexploitation of groundwater resources causes huge environmental damage - subsidence, seawater intrusion, salinity degradation, pollution, and falling groundwater levels. The economic consequences of these effects are great. The serious pollution of surface water and surface water scarcity has driven the increasing use of groundwater during the past few decades. Groundwater quality, especially in the deep aquifers, is far better than the surface water quality. However, increasingly serious pollution of the shallow aquifers is now encouraging the growing use of deep groundwater resources. The very serious and largely irreversible falling groundwater levels throughout the northern China Plain demands a major programme of groundwater management planning to reduce groundwater use to sustainable levels.

The groundwater management strategy proposed will require a huge effort; however, the consequences of not doing so will have serious long-term side-effects, such as effectively destroying the groundwater-dependent agricultural base, massive subsidence and seawater intrusion, virtual elimination of groundwater as a water source for many cities and countless households, and the loss of "insurance" water for future generations. The fundamental objective proposed is to reduce groundwater use to sustainable levels by 2015. A broad range of technical, institutional and management actions are required to achieve this goal. The key actions required are:

(a) The definition of all significant groundwater usage areas as Groundwater Management Units and the determination of the sustainable yield;

(b) Preparation and implementation of groundwater management plans as per the programme described here;

(c) Allocation licensing to be linked to sustainable yield assessment;

(d) Allocation licensing only to be undertaken by one department;

(e) Licensing of well construction drillers;

(f) A River Basin Groundwater Database needs to be developed;

(g) A groundwater pollution prevention strategy needs to be prepared;

(h) A review by the Ministry of Water Resources of the adequacy of regulations required to implement the proposed groundwater management reforms;

(i) The introduction of realistic groundwater prices; and

(j) The preparation and introduction of a major education programme on groundwater processes and the need for groundwater management.

(3) *Third Level:* The third level comprises the improvement of water resources institutional management, and integrated planning.

The improvement of water resources institutional management should involve:

(a) The establishment of a high-level River Basin Coordinating Committee or Council in the Haihe River basin, chaired by a Vice-Premier with the Ministry of Water Resources as Vice-Chairman, and comprising: (i) the mayors and governors of the constituent municipalities, provinces and autonomous regions, (ii) the Vice-Minister for Water Resources and the Vice-Ministers of SEPA and the other key "water" agencies - the Ministries of Agriculture, Construction, Finance, Communications, the National Development and Reformation Committee, and the State Meteorology Authority; (iii) the Commissioner or Director of the respective RBCM; and (iv) representatives of key civil societies such as water supply companies and water user associations;

(b) The Haihe River Water Resources Commission should be charged with the responsibility for: (i) determining water resources allocations (surface and groundwater) for the provinces (real-time, and dependent on the water held in reservoirs and seasonal conditions); (ii) the development of broad policies and programmes promoting sustainable water resources management, particularly with the respect to flood control and drought relief, groundwater management, water resources protection and pollution control, and the promotion of increased water-use (especially irrigation) efficiency; and (iii) comprehensive basin development planning;

(c) Granting the River Basin Councils the necessary legislative support and authority to ensure essential coordination and enforcement of its allocation, policy and programme decisions in the municipalities, provinces and autonomous regions;

(d) Confirmation by law of the River Basin Councils as the primary water resources management agencies in the Haihe River basin (with specific flood control responsibilities) to ensure the enforcement of council decisions on the allocations, policies and programmes in consultation with the municipal, provincial and autonomous region governments. It can be strengthened with the appropriate expertise where necessary to provide administrative and technical secretariat support for the River Basin Council; and

(e) Adoption by the River Basin Councils of the strategic planning and management methodology required to determine the policy, programme and project priorities required to accelerate the basins towards achievement of sustainable development. With regard to integrated planning, the River Basin Commission should prepare an integrated planning guide to the water resources development and protection at the province level.

(iii) Mechanisms for reporting, monitoring and evaluation

(1) *Coordination Framework:* The ministries and agencies responsible for the management, exploitation, development and use of water and water resources include:

(a) The Ministry of Water Resources. Responsible for integrated management and supervision of national water resources as well as the exploitation, development, use and protection of water resources in China. It is also responsible for, monitoring, control, promotion and reporting on the implementation of activities related to water and water resources.

(b) Ministry of Agriculture. Responsible for the exploitation, development and use of water resources in agriculture.

(c) State Forestry Authority. Responsible for the exploitation, development and use of water resources in forestry.

(d) Ministry of Communications. Responsible for the management, exploitation, development and use of water resources in the field of communications. In addition, the ministry is responsible for collection of hydrological data and hydrographical surveys for navigation.

(e) Ministry of Construction. Responsible for the management, exploitation, development and use of water resources in the municipal areas, town water supplies, urban drainage and wastewater treatment.

(f) Ministry of Public Health. Responsible for the management, exploitation, development and use of water for rural domestic consumption and health care.

(g) State Environment Protection Administration (SEPA). Responsible for ecologic environment protection and wastewater discharge management.

(h) Ministry of Land and Resources. Responsible for hydrogeology surveys.

(i) HRWRC. Empowered to exercise authority of water administration in the above-mentioned areas. Responsible for managing the water resources and river course of the local basin, according to the rule of management in unity and rank. Responsible for controlling and developing important water conservancy projects comprehensively. Ensure efficient planning, management, coordination, supervision and servicing. Undertake river control, and all-around development, utilization and water resources protection.

The above ministries and agencies would need to coordinate with the local authorities in the detailed determination of responsibilities and scope of activities within their sectors.

(2) Licensing of water use and water resources development activities

The State Council has issued Decree No. 119 on licensing water catchments. The Decree determines that the Ministry of Water Resources will implement this licensing system and the different scales of water catchment licences to be issued by water administrative institutions.

The water administrative departments should approve important medium-scale water uses. The Ministry of Water Resources or the River Basin Commissions should be responsible for approving the water catchments of individuals and organizations wishing to use water from the Yangtze, Yellow, Huaihe, Haihe, Luan, Pearl, Songhua, Liaohe, Jinsha and Han Rivers, international rivers, State border rivers, inter-province rivers indicated reaches, provincial border rivers and lakes, inter-provincial regions and water quantities beyond the set quota. The water administrative departments of local authorities should be responsible for approving other water catchments.

Permission for large-scale water resources development activities should be sought from the Ministry of Water Resources, SEPA and NDRC, and final approval given by the State Council. Applications for large-scale water resources development activities should be accompanied by a feasibility study and environmental impact assessment. Smaller-scale water resources development activities should apply to the water administrative departments and local government authorities.

The Ministry of Water Resources and SEPA need to prepare guidelines on the preparation of feasibility studies and environmental impact studies, respectively, in accordance with the Environmental Impact Evaluation Law. The ministry also needs to monitor and control the development of water resources.

All water resource development projects, particularly the construction of dams, irrigation schemes, and the diversion of rivers and streams, should go through a public consultation process with the local population. When a water resource development project is expected to have an impact on the natural or social environment, SEPA, the Ministry of Water Resources and local authorities need to:

(a) Issue a project notification to inform the public through newspapers, public notices and other means and, depending on the type of project, determine the duration for such a notification; and

(b) Organize public hearings in the relevant location. The developer or representative should attend the public hearings and provide a detailed explanation and information on the proposed mitigation measures. All costs associated with the public hearing will be the responsibility of the developer.

(3) *Monitoring and Evaluation:* The Ministry of Water Resources, HRWRC and local authorities are responsible for the monitoring and evaluation of the implementation of measures provided in relative regulations and decrees and will report to State Council on a regular basis. There are three types of monitoring and evaluation: (a) regular evaluation; (b) monitoring with prior notice; and (c) evaluating without prior notice.

(iv) Indicators

(1) A reduction in the number of water conflicts

(2) A reduction of water pollution and the improvement of water ecosystems, particularly wetland ecosystems

(3) Water productivity.

Integrated Strategic Plan for the Huaihe River Basin

Overall Social-economic Development Context of the Huaihe River Basin

The Huaihe River basin spans 35 prefectures and 181 counties in the five provinces of Hubei, Henan, Anhui, Shandong and Jiangsu. In 1997, the population was 160.43 million, with an average density of 594 persons/km2, which was 4.8 times the national average of 122 persons/km2. That made it one of the most densely populated regions in China.

The Huaihe River basin has a cultivated area of 183.26 million mu (12.22 million hectares). The main crops are wheat, rice, corn, yams, beans, cotton and rape. In 1997, the total agricultural output was RMB 284.4 billion (US$ 34.35 billion) or RMB 1,812 (US$ 218.8) per capita, which was significantly higher than the national per capita average of RMB 1,345 (US$ 162.4). These figures clearly indicate the basin's key role in national agricultural production. The main agricultural yield of the basin is grain. In 1997, the grain yield was 84.96 million metric tons, which accounted for 17.3 per cent of the total crop yield in the basin.

The Huaihe River basin is one of China's developing industrial regions. In 1995, the gross value of industrial output in the basin was RMB 763.4 billion and a GDP of RMB 541.2 billion (US$ 92.2 billion and US$ 65.4 billion, respectively). GDP per capita was only RMB 3,398 (US$ 410), which is lower than the national average of RMB 4,810 (US$ 581). The main industrial sectors in the Huaihe River basin are coal, power, food, light industry and textiles.

Opportunities and Threats for Water Resource Development from the Perspective of the Water Resources Council

Following the serious flooding in 2003, the State Council gave high priority to harnessing the Huaihe River. Investment has been greatly increased in order to guarantee the completion of important engineering construction works. The ability to prevent flood disasters is being promoted. Investment in environmental protection has also been increased, and the authority of the State Environment Agency has been strengthened with regard to controlling pollution in the Huaihe River basin. Integrated water resource management is taken into consideration during policy-making. The main challenges are flooding and water logging, water pollution and water scarcity.

A combination of climatic conditions and topography makes Huaihe River basin vulnerable to widespread, prolonged and devastating floods. While flooding event probabilities remain outside the control of society, protection against damage can be enhanced. Unfortunately, this is no easy task in the Huaihe River basin due, in part, to the climate and topography; however, an additional factor is that of the historical pattern of settlement in the plains and the extreme population density, which requires extensive infrastructure.

Surface and groundwater quality has been seriously degraded due to the lack of effective pollution control, combined with rising population and industrial operations, especially during the 1980s and 1990s. Causes of pollution from point sources include rapid urbanization, industrial development, rising population and the growth in the number of township and village enterprises and livestock operations in rural areas. Diffuse releases of nutrients and pesticides from agriculture are also becoming important sources of shallow groundwater and surface water contamination. Pollutants of concern recorded by SEPA from point sources include parameters for organics such as COD and BOD, phenols, ammonia nitrogen and inorganics such as heavy metals, while non-point source pollutant parameters include non-ionic ammonia, TP, TN and COD.

This dramatic decline in surface and groundwater quality continues despite efforts by regulatory authorities, including SEPA, and the Ministry of Water Resources, to improved control of the obvious sources of pollution such as state-owned enterprises as well as attempts to close small highly-polluting town and village enterprises. The current surface water quality status is such that most rivers and lakes fail to meet the standards required for the designated beneficial uses of the water. In 2002, the river stretches with water quality equal to, or worse than class IV was 69 per cent of the total river length.

In general, the water shortage problem in Huaihe basin is not particularly serious. However, despite the region's monsoon climate, with its greater variability in precipitation, water shortage problems occur in some dry years. Therefore, as a result of socio-economic development combined with water pollution, water shortage problems will become serious in the future.

Vision Statement of the Huaihe River Water Resources Commission, Ministry of Water Resources

This Vision Statement expresses the long-term objectives of the water sector. HRWRC, under the Ministry of Water Resources, represents the ministry in the management of water resources in the Huaihe River basin. As proposed by working group members for the Huaihe sector, the main vision statement of HRWRC states: "To provide adequate safety to the people and economic production systems by 2010, recover the main river ecosystems for people and aquatic life by 2020 and provide a sustainable water supply to support sustainable socio-economic development of the Huaihe River basin".

Some of the reasons and thinking behind this statement are worth elaborating. Flood safety refers to the establishment of a comprehensive system to prevent flood disasters. Water supply safety means having enough clean water to supply to different water-use sectors. Ecology safety refers to water development that is harmonious with the ecology. All three aspects guarantee the sustainable development of the basin.

Legal and Institutional Framework of the Huaihe River Basin

The Huaihe River basin has no local laws. All laws in China concerning water, such as water law, flood prevention law, water and soil conservation law, among others, will protect and control river basin water resources development and conservation. According to the local situation, local government and river basin commissions can formulate some ordinance.

China has long focused on flood control and irrigation, but in the past 50 years the problems of urban and industrial water supply, water pollution control and protection of aquatic ecological resources have also become major issues.

The Government now recognizes the great importance of making the best possible use of available water in non-flood periods as well as the continuing importance of better flood management. It also recognizes that in order to achieve this objective, a marked change is required in the existing institutional pattern related to water resources management, which is now highly fragmented between national, basins and provincial local interests and agencies. To do this, a new institutional system needs to be established that recognizes that the salient approach must be based on planning where the entire river basin is the controlling, unit, with the national and provincial local governments working in coordination to achieve optional water utilization within the basin. This would include both surface water and groundwater.

The Ministry of Water Resources has the primary responsibility for overall management and development of the nation's water resources and is well skilled and qualified to address the myriad problems. However, in an operational sense, its specific functions for flood protection and water management are implemented through the seven established River Basin Commissions.

In addition, the ministry's authority is limited in many critical areas, given the high degree of autonomy of China's provinces and overlapping jurisdiction of other ministries and agencies, particularly in the areas of urban water supply, groundwater management, pollution control and operation of reservoirs for hydropower. The result is that the current water management system is somewhat confused, not cohesive, and fraught with opportunities for over-allocation, which has been confirmed by the grossly depleted groundwater resources.

The River Basin Commissions are, in general, weak compared with the economic and political strength of the provinces. Institutional fragmentation from the central government downward explains a large part of this, since it does not allow planning and day-to-day operational management to be effectively realized over the entire river basin. It is suggested, therefore, that the provinces' pre-eminence over natural resources management and development within their borders, should be formally recognized, subject to "technical" oversight by the respective river basin commissions, particularly their responsibility for the management and use of their respective shares of interprovincial surface and groundwaters.

At the river basin level, the institutional challenge is to bring the provinces together with the key "water" agencies, so that they all have a direct role in governance to ensure sustainable development of the river basin as a whole.

The mission of Huaihe River Basin Commission includes the following aspects:

(a) On the basis of strengthening construction measures, also strengthen non-construction measures to promote synthesis capability in flood prevention and disaster reduction. It is planned to treat the river course, open up new floodways and strengthen the levees on the both banks. Considering the use of detention areas is one of the most important components of the flood control project system in the Huaihe River basin. There are nine detention areas in the basin with a total storage capacity of 8.61 billion m3;

(b) Guarantee clean water supplies for urban and town areas, not only for the domestic and industrial sectors but also to ecological water. As for agriculture in the basin, consideration should be given to different water sources unification, while irrigation guarantee probability should not be increased; and

(c) Take effective measures to save water and control pollution. Water saving and water pollution control are preconditions of water resources development. It should promote Water-use efficiency should be promoted in agriculture. By water saving, discharges of sewage water from urban areas will be reduced. Encourage the industrial sector to implement cleansing production, and the control of pollution from polluting sources.

Strategic Plan Goals

Integrated water resources management in the Huaihe River basin should coordinate many sectors, such as those concerning water resources, construction, water-borne transportation, ecology protection, energy development, water development and water pollution control, among others. Water resource management should focus on the basin level and include unified basin planning. With integrated water resources management, the Huaihe River Basin Commission will be able to achieve the general goal, i.e., sustainable water resources use in support of sustainable basin development. Within the context of this overall goal, the Commission will attempt to achieve the following three main mission goals.

a. IWRM Goal 1

Establish a comprehensive flood prevention system, clean water supply system and water protection system. To achieve this goal, the following legal and institutional measures will be taken:

(1) *Improved implementation of the Law on Water and Water Resources:* Formulate a new water law that emphasizes water basin management and give the River Water Resource Commission legal status. The Ministry of Water Resources will work in cooperation with all key ministries and stakeholders towards effectively managing, developing and protecting water and related resources for agricultural, municipal, industrial, rural, hydropower, recreational, and fish and wildlife purposes. The Ministry of Water Resources and the Huaihe River Basin Commission will pursue a coordinating role aimed at improving water and related resource management, development and protection activities. The framework for such a coordinating role will include:

   (i) Introducing legislation to improve integrated and comprehensive water resources management, including flood control, irrigation, urban and rural water supply, water pollution control and treatment etc.; and

   (ii) Introducing legislation to strengthen the authority of the River Basin Commission to implement integrated and comprehensive water resources management. The River Basin Commission should be given more implementation authority to ensure sustainable water use, not only design and plan jobs.

(2) *Improvement of the Flood Prevention Planning System:* Flood prevention planning has been undertaken in China, but with advances in socioeconomic and scientific

development, flood prevention planning should be modified. It should mainly emphasize harmonization with flooding and the use of non-construction measure.

(3) *Improvement of the Water Resources Planning System:* China is now formulating comprehensive water resources planning, including surface water and groundwater, water quantity and quality, water use and protection, economic water user and environmental water users etc. The new water resources plan should receive approval from the State Council and should provide guidelines for management activities. After several years, water resource planning should be amended according to the changed conditions.

(4) *Promotion of Applied Sciences and Technology Integration:* New concepts and new technology should be used in water resources management, water saving, water use and water protection.

b. IWRM Goal 2

Strengthen administration and make full use of market mechanisms to enhance and consolidate existing systems and the foundation to operate, maintain and rehabilitate facilities safely, reliably and efficiently in order to protect public investment.

Under this mission goal, HRWRC will focus on operating, maintaining and rehabilitating existing water resources facilities, so that they continue to provide project benefits. These facilities provide power and water supply delivery systems that serve agricultural and municipal users, in addition to recreation, fish and wildlife benefits, and flood control. By ensuring that facilities are safe, cost-effective and reliable, HRWRC can assist the agencies concerned to operate them so that they provide project benefits while protecting public health, sustaining environmental values, and providing timely and economical service to customers and, thus, protect the public investment.

For this mission, the related agencies will be assisted in conducting oversight reviews as well as planning to ensure timely operation and management procedures, including control and communications systems for the continued safe and reliable operation of existing facilities. In this connection, it is expected that the related agencies will be able to verify reliability, improve maintenance practices, and identify and implement risk reduction actions.

Ensure Effective Operations of Facilities

It is important that related agencies comprehensively review, update and implement Standard Operating Procedures and related operating guidance to ensure that existing facilities provide project benefits in a safe and reliable manner.

(i) *Strategies for Achieving this Goal:* HRWRC will, through the implementation of related laws and development of regulations, encourage all operating agencies to use appropriate tools, technologies, innovations and research to operate facilities effectively, including internal auditing systems.

(ii) *Programme Evaluation and Performance Improvements:* Establishment of indicators and monitoring of incidents will be necessary for effective evaluation of performance improvements.

(iii) *Performance Measures:* Report on emerging issues in the operation of existing facilities in the water sector in 2005, and propose a strategy for improving the efficiency and effectiveness of existing facilities in 2006.

Effective Participation by the Public

Participation of the public is essential in enhancing the benefits of existing facilities and minimizing the risks and impacts of disasters. In addition, effective public participation will result in better investment strategies in terms of resources mobilization and investment scheduling.

(i) *Strategies for Achieving this Goal:* The following approach is necessary. First, develop an awareness and consultation strategy that will (a) identify target groups and their information needs, appropriate and cost-effective means of providing information and creating awareness,

(b) identify modalities and an action plan for consultation with stakeholders regarding draft policy and legislation, and other important consultation topics, and (c) identify awareness and consultation roles of central government agencies and HRBC in relation to all major target groups.

The next steps are to carry out awareness and consultation responsibilities according to the agreed strategy, play an awareness and consultation role at the river basin level, and develop and implement national guidelines for dam safety.

(ii) *Cross-cutting Relationships:* As the overall coordinating body in water resources management, HRWRC will assist other agencies in the development of the above strategies and guidelines, including establishment of an appropriate mechanism for the implementation of the strategies and guidelines. Programme evaluation and performance improvements: Detailed indicators will be developed to monitor the implementation of the related strategies and guidelines.

(iii) *Performance Measures:* Establish working groups on public awareness and participation.

c. IWRM Goal 3

The goal is to enhance the organizational capacity and effectiveness of the water resources coordination system. HRWRC needs to be strengthened to make it an effective organization and enable it to achieve the above mission goals. Under this mission, the service provided to clients will be improved, particularly for top decision-makers in the water sector and the Government.

In addition, its leadership role in the coordination of water resources management, development and protection will be strengthened, and support will be provided for an innovative and growing workforce able to respond to diverse needs. To improve the business approach, it is essential to involve decision-makers and employees of HRWRC in the formulation and implementation of this strategic plan and the subsequent monitoring and evaluation.

Strategies for Achieving this Goal

(a) Preparation of an Action Plan for two- and five-year periods. The Action Plan will be based on the mandate of HRWRC as well as priority water resource management issues at the national level and in the river basin. It will be necessary for the Action Plan to emphasize the need to integrate a more balanced view of sectoral perspectives in the programmes of investment and operations of all water-related agencies. HRWRC will present the Action Plan to the Government for approval.

(b) Prepare and implement a capacity-building plan for HRWRC, based on the requirements for achieving the Action Plan.

(c) Conduct training needs assessment for IWRM training in HRWRC and the provinces in the river basin. Prepare a coordinated training plan to address the priority training needs, drawing on the training modalities identified in the project and the available training budget. Present the training needs assessment and plan to HRWRC for approval.

Implementation, Monitoring and Evaluation

Having established a strategic plan management process, the various aspects of implementation are considered below.

(1) Human Resources Development

In developing human resources, it will be necessary to align the organization and develop the necessary light skills and competencies to match the strategy. It is vital to recognize that people are the best source of ideas for the future, and that they form the foundation for creating good leadership and teamwork as well as the organizational culture needed for the successful implementation of strategic plan management. Thus, the following steps need to be taken:

- Build the organization around the critical strategic elements
- Mobilize and allocate resources to match the strategic plan

- Establish administrative systems, policies and procedures that are in accordance with the strategic plan
- Develop a personnel rewards system linked to achievement and the strategy plan
- Recognize that the most important job of the leadership is to establish and have confidence in the strategic plan management and to ensure that everyone in the HRWRC follows it

(2) Funding for the Future

Because of the complexity involved in strategic planning, it is necessary to acquire financial support both at home and abroad. Only if HRWRC treats the raising of funds as a fundamental part of the operation, and approaches it in a professional and systematic way, will strategic planning have any chance of becoming financially secure in the future. Since the Ministry of Water Resources has appointed HRWRC as the key leading institution in the promotion and coordination of water and water resources development for the benefit of the river basin's population, it is unquestionably the duty of HRWRC to treat fund raising as an essential component of the overall operation, and not just as an auxiliary issue.

Financial security means that funding must be sought and obtained in a flexible way in order to take the advantage of diverse sources, including the central Government and the Ministry of Water Resources in particular, and local government, communities and associations, involved enterprises, international organizations and so on. It is certain that the central Government subsidy will play the most crucial role. Local government contribution will be in the form of counterpart funds. Donations contributed by other institutions or organizations will take many forms, such as cooperative scientific research, transfers of the project outcome at a preferential price or even free of charge etc. As for bursaries from abroad, these can be provided in similar forms as above. In addition, HRWRC is responsible for identifying a reasonable goal for funding and in-kind services during the process of strategic implementation, especially with regard to what these funds and services should be used for, and how long it may take to reach this level of contribution. With regard to financial management, HRWRC is accepted as the core agency in charge of relevant affairs. Each of its functions is explicitly defined by the central Government, and in particular the Ministry of Water Resources, and supported by local governments at the different levels. The major duties to be performed by HRWRC are:

(a) Putting forward the overall budget
(b) Detailing the amount of funding needed to cover all projected expenditures
(c) Determining the organization of fund-raising in order to clarify the contribution percentage from each financial resource for the foreseeable future
(d) Establishing and implementing efficient financial monitoring systems for the strategic planning period

(3) Indicators

The indicators include:

(a) The decrease in the number of water conflicts
(b) The reduction in water pollution and improvement in water ecosystems, particularly wetland ecosystems
(c) Water productivity

Action Plan: The River Basin Commissions will undertake activities that are categorized in three levels according to priority.

First Level

(i) *Coordination:* The River Basin Commissions need to strengthen coordination with the provinces concerned and between themselves.
(ii) *Information:* Establish a consolidated information system. This will require the provinces to provide basic hydrological and water use data to the River Basin Commissions, while the River Basin Commissions should provide information to the provinces and the public. This consolidated information system will not only reduce water-use conflicts but also improve water-use efficiency.

(iii) *Reasonable Allocation of Water Resources:* Water resources include surface water, groundwater, treated wastewater, brackish water and freshened seawater. Water use involves many aspects, such as domestic urban and rural usage, production usage and eco-environment usage. Optimized water allocation will be achieved via water allocations based on time and area, coordination of water supply and usage, unification of water quantity and quality, real time regulation and preference of benefits. The mutual relationship of economic structures, economic development and water resources must be handled correctly via administrative, economic, technical and legal measures. Reasonable water resources allocation can ensure that limited water resources produce full benefits.

(iv) *Water Pollution Control:* The Huaihe River system was heavily polluted due to population growth and the increase in the number of township and village enterprise, without much attention being given to sewage and wastewater treatment. The River Basin Commissions will be granted authority to coordinate with the Environment Agency in taking effective measures to control point source and non-point source pollution.

Second Level

(i) Construction and non-construction measures for flood control and water-logging

This level involves three main actions:

(a) Reinforcement of the reservoirs along the Huaihe River. There are 5,100 reservoirs in the mountainous/hilly region of the Huaihe River basin, of which 35 are large-scale reservoirs with a total storage capacity of 17.86 billion m3 and a controlled area of 32,300 km2. They account for 40 per cent of the total mountainous region. However, most reservoirs were built during the 1960s when the "Great Leap Forward" occurred and design and construction standards were sacrificed in the interests of expediency. Dangerous reservoirs should be adapted to ensure that the reservoirs play a greater role in flood control in a safe way;

(b) The use of detention areas. This is one of the most important components of the flood control project system in the Huaihe River basin. There are nine detention areas in the basin with a total storage capacity of 8.61 billion m3. Among them, the Mengwa, Chengxihu, Chengdonghu and Wabuhu detention areas are located in the middle reaches of the Huaihe River and have a total storage capacity of 6.57 billion m3. They can reduce the flood discharge by 20 per cent and weaken peak flow very effectively. There are 19 floodway areas; these can be used to increase the discharge to the downstream stretch of the Huaihe River. An analysis indicates that floodways can divert discharge by 30 to 50 per cent during minor flooding and by 50 to 70 per cent during serious flooding; and

(c) Improve the levee flood standard on the Huaihe River. It is planned to improve the river course and open up new floodways to strengthen the levees on both banks. After flooding is regulated by Hongze Lake in the upper and middle reaches of the river, the normal flood outlets in the lower reaches are first into the Yangtze River (mainly as a flood diversion outlet) and then into the sea through the general irrigation channel. In addition, when Huaihe River floodwater does not meet Yishusi floodwater, the former can enter the sea through the New Yi River after being diverted by the Huaishu River. When a mega flood occurs, all the above mentioned strategies will have to be employed. If an emergency occurs at the Hongze Lake and Lixia River levees, water has to be temporarily diverted from the north of the general irrigation channel.

(ii) Water resources projects

This aspect involves:

(a) Establishing the three cross-basin water transfer projects - the east and middle south-to-north water transfer projects and the diversion from the Yangtze River to the Huaihe River are cross-basin water transfer projects. The eastern south-to-north water transfer project can solve the water scarcity problem in areas along the Great Channel in Jiangsu, Anhui and Shandong Provinces. The middle south-to-north water transfer project can solve the problem along the upper reaches of the Huaihe River in Henan Province. The diversion project from the Yangtze River to the Huaihe River can solve the problem for the Huibai area;

(b) Construction of water diversion and bailing projects for using local water resources. Henan Province plans to construct a water diversion project with a discharge capacity of up to 23 m^3/second. Anhui Province plans to construct the Wuhe-Huaihe diversion to irrigate 110,000 hm^2;

(c) Construction of urban and town water supply projects.

(iii) Irrigation and water-saving projects

In the Huaihe River basin, irrigation and water-saving projects mainly include large rebuilt irrigated areas, and the alteration of middle and lower production farm areas. During the tenth water resources planning exercise, in order to optimize of water allocation and integrated management, the Huaihe Water Resources Commission should pay close attention to strengthening the management of irrigated area construction.

(iv) Water and soil conservation projects

According to the outlines of national water and soil conservation planning and the ecology construction planning, the Huaihe River basin will focus on: (a) establishing a monitoring network as well as use small watersheds as units to ensure comprehensive harnessing; (b) strengthen water resources protection; (c) introducing water resource protection planning and water resource function zone planning; (d) establishing a water resource protection information system; and (e) establishing a water source protection zone in the city.

Third Level

(i) Improvement of water resources institutional management

The improvement of water resources institutional management will involve:

(a) The establishment of a high-level River Basin Coordinating Committee or Council (RBCC) for the Huaihe River basin, chaired by a vice-premier from the Ministry of Water Resources as vice-chairman. The members would comprise the mayors and governors of the constituent municipalities, provinces and autonomous regions, the vice-minister for water resources and the vice-ministers of SEPA and the other key "water" ministries - agriculture, construction, finance and communications - as well as the National Development and Reformation Committee, the State Meteorology Authority, the Commissioner or Director of the respective RBCC, and representatives of key civil societies such as water supply companies and water user associations;

(b) Mandating the Huaihe River Water Resources Commission with responsibility for (i) the determination of water resources allocations (surface and groundwater) for the provinces (real-time and dependent on the water held in reservoirs and seasonal conditions), (ii) the development of broad policies and programmes promoting sustainable water resources management, particularly with regard to flood control and drought relief, (iii) groundwater management, water resources protection and pollution control, and (iv) the promotion of increased water use (especially irrigation) efficiency; and (v) comprehensive basin development planning;

(c) Providing the River Basin Council with the necessary legislative support and authority to ensure the essential coordination and enforcement of their allocation, policy and programme decisions in the municipalities, provinces and autonomous regions;

(d) Confirmation by law of the River Basin Council as the primary water resources management agency in the Huaihe River basin (with specific flood control responsibilities), to ensure the enforcement of council decisions on the allocations, policies and programmes in consultation with the municipal, provincial and autonomous region governments. This will be strengthened with the appropriate expertise, where necessary, for providing administrative and technical secretariat support for the River Basin Council; and

(e) Adoption by the River Basin Council of strategic planning and management methodology to determine the policy, programmes and project priorities required to accelerate the basin towards achievement of sustainable development.

(ii) Mechanisms for reporting, monitoring and evaluation

The coordination framework comprises ministries and agencies responsible for the management, exploitation, development and use of water and water resources. These ministries and agencies include:

(a) The Ministry of Water Resources. Responsible for the integrated management and supervision of national water resources, the exploitation, development, use and protection of water resources in China, and monitoring, control, promotion and reporting on the implementation of activities related to water and water resources.

(b) The Ministry of Agriculture. Responsible for the exploitation, development and use of water resources in agriculture.

(c) The State Forestry Authority. Responsible for the exploitation, development and use of water resources in forestry.

(d) The Ministry of Communications. Responsible for the management, exploitation, development and use of water resources in the field of communications. In addition, the ministry is responsible for the collection of hydrological data and hydrographical surveys for navigation.

(e) Ministry of Construction. Responsible for the management, exploitation, development and use of water resources in the municipal field - town water supply, urban drainage and wastewater treatment.

(f) The Ministry of Public Health. Responsible for the management, exploitation, development and use of water for rural domestic consumption and health care.

(g) The State Environment Protection Administration. Responsible for ecologic environmental protection and wastewater discharge management.

(h) The Ministry of Land and Resources. Responsible for hydrogeological surveys.

(i) The Huaihe Water Resources Commission (HWRC), which is empowered to exercise authority on water administration in the above-mentioned areas, will (i) manage the water resources and river course in the local basin, (ii) be responsible for controlling and developing important water conservancy projects comprehensively, (iii) ensure efficient and effective planning, management, coordination, supervision and servicing, and (iv) improve river control as well as all-around water resources development, utilization and protection.

The above-listed ministries and agencies will coordinate their efforts with the local authorities in order to determination details of responsibilities and the scope of activities within their sectors.

(iii) Licensing of water use, and water resources development activities

The State Council's Decree No. 119 on licensing water catchments authorizes the Ministry of Water Resources to implement this licensing system. Water catchment licences are issued by determined the scale of the water administrative institutions concerned. The water administrative departments should approve important medium-scale water uses. The Ministry of Water Resources or the River Basin Commissions should be responsible for approving the water catchments of individuals and organizations wishing to use water from the Yangtze, Yellow, Huaihe, Haihe, Luan, Pearl, Songhua, Liaohe, Jinsha and Han Rivers, international rivers, State border rivers, inter-province rivers indicated reaches, provincial border rivers and lakes, inter-provincial regions and water quantities beyond the set quota. The water administrative departments of local authorities should be responsible for approving other water catchments.

Permission for large-scale water resources development activities should be sought from the Ministry of Water Resources, SEPA and NDRC, and final approval given by the State Council. Applications for large-scale water resources development activities should be accompanied by a feasibility study and environmental impact assessment. Smaller-scale water resources development activities should apply to the water administrative departments and local government authorities.

The Ministry of Water Resources and SEPA need to prepare guidelines on the preparation of feasibility studies and environmental impact studies, respectively, in accordance with the Environmental Impact Evaluation Law. The ministry also needs to monitor and control the development of water resources.

All water resource development projects, particularly the construction of dams, irrigation schemes, and the diversion of rivers and streams, should go through a public consultation process with the local population. When a water resource development project is expected to have an impact on the natural or social environment, SEPA, the Ministry of Water Resources and local authorities need to:

(a) Issue a project notification to inform the public through newspapers, public notices and other means and, depending on the type of project, determine the duration for such a notification; and

(b) Organize public hearings in the relevant location. The developer or representative should attend the public hearings and provide a detailed explanation and information on the proposed mitigation measures. All costs associated with the public hearing will be the responsibility of the developer.

(iv) Monitoring and evaluation

The Ministry of Water Resources, HRWRC and local authorities are responsible for the monitoring and evaluation of the implementation of measures provided in relative regulations and decrees and will report to State Council on a regular basis.

Three types of monitoring and evaluation are required:

(a) The number of water conflicts, such as those concerning the allocation of water and pollution transfer, will provide the statistics and constitute the most important indicators for evaluating the action plan;

(b) Water quality, expressed by water class, will be an important indicator for monitoring the progress of water pollution control; and

(c) Water supply safety and water productivity will be evaluated to test the efficiency of water projects and the ability of water systems to support economic growth.

Overall Socio-economic Development Context of Priority Areas and Priority Socio-economic Goals

China's GDP has risen substantially since reform and opening of the market. It was earlier projected by the National Development and Reform Commission that GDP would increase at the rate of 7 per cent annually during 2001-2005, and that the average GDP per capita would be RMB 9,400 in 2005. The population is projected to reach 1.33 billion in 2005. Social and economic development will place much greater pressure on the water sector.

Opportunities and Threats Facing Water Resources Development from the Perspective of the Ministry of Water Resources

Since the great flood in 1998, the Government has been paying much closer attention to integrated water resources management, which is mainly reflecting in the following areas. The status of the water sector as an infrastructure and base domain has strengthened in China. Investments in the water sector by both he central and local governments have increased. In addition, a new facet - the water sector for society and society support for the water sector - has been formed in principle.

Social and economic development provides good conditions for construction in the water sector. With socio-economic development and the increase in individual financial capacity, the demand being placed on the water sector is also increasing. After the entry of China into the World Trade Organization, the country will experience an inflow of advance technology, overseas capital and new concepts of water management, giving many advantages and opportunities to the water sector. In addition, water savings among the public have strengthened.

Although great progress has been made, serious challenges still face the development of the water sector, particularly in the areas detailed below.

The gap between water supply and demand in China has become steadily wider. Currently, the annual agricultural water shortage in the whole country is about 30 billion m^3, and the annual average total area of farmland affected by drought is 20 million hm^2, resulting in a grain yield reduction of between 10 billion kg and 25 billion kg, or even as high as 50 billion kg in years of serious drought.

The annual shortage of urban and industrial water is about 6 billion m^3, resulting in an industrial output reduction of RMB 230 billion. Of the 660 cities in the country, more than 400 suffering water shortages, affecting 160 million people. In addition, more than 24 million rural inhabitants are without access to an adequate drinking water supply. In particular, in the plain areas in the lower reaches of the Yellow, Huaihe and Haihe Rivers, the average per capita water use is only 332 m3 while the annual water shortage is up to 18 billion m3, ranking them among those with the most serious water shortages in the whole country.

Serious threats of flooding still exist in China. The country's flood control system is still inadequate, with low standards and low capacity for flood control. Most of the current embankments along the major rivers only meet a flood prevention standard of a 10-year to 20-year flood. Of the cities at and above the prefecture level, only 17 per cent reach the originally defined flood prevention standard. The development of flood storage and detention zones has lagged, resulting in problems in their application. Since the 1990s, in particular, extraordinary floods have continuously occurred along the major rivers, causing a total economic loss of

RMB 1,000 billion. The floods that occurred along the Yangtze and Songhua Rivers in 1998 alone caused direct economic losses of more than RMB 250 billion. Thus, flood control remains an arduous task for China.

Water-use efficiency in all water-use sectors remains low. In the past 20 years or more, China has made great efforts to promote water-use efficiency and has achieved significant progress; however, a big gap remains compared with the advanced levels in other parts of the world. For example, in 2000, the national integrated irrigation water efficiency was 0.43, but the advanced irrigation water efficiency of foreign countries was above 0.7. In the same year, the rate of industrial water reuse in China was 0.55, while in the advanced countries it was between 0.75 and 0.85. As agriculture accounts for 70 per cent of the total water use in China and tertiary industries comprise only a small proportion of all industries, the GDP per unit volume of water use in China is even farther behind the advanced world level.

Water pollution is heavy and has basically not been controlled. The natural quality of river flows in China is generally good, but the rivers have been polluted to a varying extent due to human activities and increasing discharges of domestic, industrial and agricultural wastes. According to recent statistics, the total discharge of sewage and wastewater was 62 billion tons in 2000, of which only 24 per cent was treated collectively up to the national standard. The remainder was either not treated or treated but not up to the national standard before being discharged or used for farmland irrigation.

In the 100,000 km of river reaches that were evaluated throughout the country, water in the category of grade IV or worse accounted for 47 per cent; while more than 75 per cent of the lakes were polluted. An investigation of drinking water in 118 cities indicated that groundwater had been polluted to a varying degree in 97 per cent of the cities, of which 64 per cent was serious pollution. Water pollution has caused ecological deterioration, declines in water quality in rivers, lakes and reservoirs, and has reduced the value of useable water. This has made water shortages even more serious.

Soil erosion and ecological deterioration are serious in some parts of the country. China is among those countries with the most serious soil erosion. In addition to natural factors, human activities have also increased soil erosion. Since the founding of the People's Republic of China, the Chinese Government has invested a great deal of resources in soil erosion control, but the trend of worsening soil erosion and ecology has not been checked. According to the second national survey on soil erosion, based on remote sensing, the current area of soil erosion in the whole country is 3.56 million km^2, accounting for 37 per cent of the national land area. In some areas, particularly the western regions, ecological deterioration has been virtually unaltered.

At the same time, the phenomenon of damaging activities coexisting with remedial efforts still remains. China has a large number of wetlands. However, the pressure of population growth and economic development has continued to damage the wetlands to varying extents. In the past few decades, the total areas of lakes in the country have decreased by 1.3 million hm^2 and, on average, 20 lakes have disappeared each year. The degree of river flow exploitation is more than 50 per cent for most of the major rivers, thus resulting in silting up of river channels or mouths, drying up of middle or lower reaches and of lakes in lower reaches, the reduction or disappearance of natural vegetation, desertification and serious damage to aquatic life. Groundwater overdraft has created 72 depression cones around the country, covering 61,000 km^2, hence resulting in ground subsidence, cracking and other geological hazards or environmental problems.

Vision Statement of the Ministry of Water Resources

In order to carry out the new development perspective of the Central Committee of the Communist Party of China, the Ministry of Water Resources has changed its thinking on implementing sustainable approaches to water resources management. The essence of the new viewpoint is harmony between humans and nature, requiring changes in traditional ideas and behaviour.

The first change is to shift from the concept that water cannot be exhausted to the concept that freshwater is a limited and vulnerable resource; from the view that water is a free natural resource to that of water being an important strategic economic resource.

The second change is a shift from unbounded development and use of water resources, without consideration being given to ecological needs, to paying careful attention to ecological use and rationally arranged domestic, productive and ecological water use.

The third change is from simply emphasizing flood control and prevention to a coordinated relationship between humans and nature in order to preserve the necessary space for flood flows and storage, and to make use of floodwater as much as possible with the prerequisite of ensuring safety against flooding.

The fourth change is a shift from biased emphasis on development, utilization and harnessing of water to coordinated development, utilization, harnessing, allocation, conservation and protection, with the focus gradually being moved to rational allocation, high-efficiency utilization and effective protection.

The fifth change is a shift from the separation between water resources and the economy and society, to basing population growth, economic structure and layout of productivity, development scale and speed on the carrying capacity of water resources, while seeking the coordinated development of population, resources, economy, society and environment.

The sixth change is a shift from only depending on engineering measures to integrating engineering and legal, administrative, economic and technical measures, and from engineering planning to strategic integrated water resources planning.

The seventh change is a shift from only using surface water or groundwater resources to jointly using surface water, groundwater, rainwater, floodwater, low-quality water, reclaimed wastewater, seawater, desalinized seawater etc.

The eighth change is a shift from water supply management that only pays attention to satisfying demand to management that applies multiple measures to rationally check demand. Establishing a water-saving society is a new task of the Ministry of Water Resources. The ministry will widely publicise water saving in society, encourage the agricultural and industrial sectors to adopt new methods of saving water, and promote the use of new equipment for domestic water saving.

Combined with environmental considerations, the Ministry of Water Resources will strengthen water resource protection in its efforts to control water pollution.

The ministry will also promote efficient water resources use further through water allocation. It is currently compiling comprehensive water resources planning, which will fully consider water usage in the various watersheds and by water use departments and the domestic sector as well as in production and the ecology.

In order to promote water-use efficiency; the Ministry of Water Resources will take into full consider the use of market mechanisms in the future. At present, water rights and water markets are not in good form; therefore, the ministry will develop them and introduce them into water resources allocation.

Mission Statement of the Ministry of Water Resources

The Ministry of Water Resources is the department of the State Council that is responsible for water administration. In accordance with the stipulations of the State Council, the Ministry of Water Resources has been given the following mandates.

(a) Formulation of water-related policies, development strategies and medium-term and long-term development plans, including water conservation and demand management policies;

(b) Drafting and implementation of enabling legislation and the design of water-related regulatory frameworks. This includes the implementation of the water-drawing permit system and the water resource fee system, and the mediation and arbitration of intersectoral and interprovincial water disputes;

(c) Implementation of integrated water resources management, including atmospheric water, surface water and groundwater. This includes:

   (i) The formulation of national and interprovincial development plans for water supply and demand, and schemes for water allocation

   (ii) Supervision of the execution of the above plans and schemes

   (iii) The assessment of water resources, flood risk and flood mitigation measures in relation to the overall planning of the national economy, urban planning and major construction projects

   (iv) The publication to national research and development in hydrology

   (v) Guidance of national research and development in hydrology

(d) The formulation of water resource protection plans in accordance with related national laws, regulations and standards concerning resource and environment protection; the demarcation of functional water areas and the control of wastewater discharges into potable water areas and other water areas; monitoring the quantity and quality of water in rivers, lakes and reservoirs; and review and approve the pollution loading capacities of water bodies with proposed limits on total wastewater discharge;

(e) The formulation of economic regulatory measures for the water sector; the enforcement of macroeconomic regulations on the utilization of funds within the water industry; the provision of guidance on economic activities related to water supply, hydropower and diversified development within the water sector; and the provision of recommendations on economic regulation of water pricing, taxation, credit and financial affairs;

(f) Drafting and reviewing of proposals and feasibility study reports on large and medium-sized capital construction projects in the water sector;

(g) Drafting and supervision of the execution of technical standards for the water sector as well as specifications and codes for water works; the implementation of key hydro-science research projects, and the popularization and dissemination of water-related technologies;

(h) The organization and direction of management protection for hydraulic facilities, water areas, dykes and coast lines, and the regulation, reclamation and development of major rivers and lakes as well as beaches; handling foreign affairs in relation to international rivers between China and its neighbouring countries; the organization of construction and management of key controlling and interprovincial hydro-projects; and the organization and direction of monitoring and management of the safety of reservoirs and dams of hydropower stations;

(i) Provision of guidance on activities related to rural water resources; the organization and coordination of capital construction of farmland drainage and irrigation, rural electrification, and water supply for townships and villages;

(j) The organization of water and soil conservation nationwide, including the formulation and development of engineering measures for water and soil conservation; and organization of monitoring and overall prevention and control of soil and water losses;

(k) Responsibility for activities concerning science, technology and foreign affairs that are related to water resources, including the provision of guidance in the development of a competent workforce for the water sector;

(l) Responsibility for the day-to-day work of the State Flood Control and Drought Relief Headquarters. This includes the organization, coordination, supervision and direction of nationwide flood control as well as the execution of flood control and drought prevention operations in major river basins and key water projects; and

(m) Other duties and responsibilities assigned by the State Council.

CONSULTATION MECHANISMS AND INITIAL FINDINGS

Consultation Mechanisms in Place and Future Developments

a. Framework for Participation

Based on the Constitution, the public participation issues have been provided for in a number of laws, decrees, notices, guidelines etc., with regard to sector-based specific requirements. Some of the specific public participation issues related to each sector are detailed below:

(a) The Environmental Law states that all persons and organizations residing in China have an obligation to protect the environment and face penalties for behaviour that is harmful to the environment. The improvement of scientific education concerning environmental protection, the development of science and technology for environmental protection, and the popularization of environmental protection knowledge are encouraged. In the institutional arrangement framework, the environment management and monitoring processes include all organizations, from the grassroots to top central levels, including village administrations, environmental management and monitoring units at the district, municipal, provincial and ministerial levels, and State environmental protection administration as a whole.

(b) The Environment Impact Evaluation Law requires that the State involves the relevant institutions, experts and the public in environmental impact assessment. Whenever planning has the potential to create a negative impact on the environment, the department in charge of special planning must organize a workshop or other meeting to enable the relevant institutions, experts and the public to voice an opinion on the draft report, which should specify why such opinions should be adopted or rejected before it is submitted for approval. The municipal and provincial level authorities of the involved district must recommend to the appointed environmental administrative department or other department that relevant representatives and experts be allowed to establish a panel to examine the environmental impact assessment report. The panel must put forward a conclusion in written form.

In cases where planning may have a major impact on the environment, the responsible department should follow up the environmental impact assessment and submit the results to the approving department. If a serious negative impact appears likely, the corresponding remedial measures should be put forward immediately. The operator responsible for carrying out such projects must then organize a workshop or other meeting to enable the relevant institutions, experts and the public to express their opinions prior to the submission of the environmental impact evaluation report for approval, except in the case of classified projects. The report should specify why such opinions should be adopted or rejected as well as the obligations of different partners, such as the project developers, and other stakeholders before it is submitted for approval.

(c) The Water Law states that water resources are national property. The developers of water resources must contribute funding for such projects and, where human resettlement is necessary, they must provide adequate arrangements and funding for such resettlement. In addition, within the framework of the Ministry of Water Resources and River Basin Commissions, some relevant key documents have been drafted, including:

(i) a draft decree to implement the Water Law;

(ii) a draft policy on the water resources development objectives, which are to ensure that the management, exploitation, use and development of water and water resources are sustainable, equitable and in support of the goals of socio-economic development and environmental protection in China; and

(iii) to organize a public awareness programme to increase the understanding of the importance of water resources, the vulnerability of these resources and clear ideas concerning the development, management, use and preservation of water resources. At the same time, the public is to be allowed to participate in the implementation in an effective manner. In the drafting process of those documents, a number of broad discussion meetings with different line agencies are to be organized at all levels throughout country, which gives the relevant agencies the opportunity to contribute their views and raise their concerns regarding the draft documents. The other objective is to gather comments, suggestions, data and information from those meetings, in order to ensure that the documents are consistent with the realistic situation prevailing in China.

b. Level of Public Participation

In China, public participation is handled at four levels:

(a) Information gathering. This involves the systematic analysis of existing social, cultural and economic conditions among the directly affected groups of stakeholders (such as farmers or indigenous minorities);

(b) Information dissemination. This involves the provision of information about a project to all interested parties (stakeholders). Stakeholders cannot genuinely be consulted or participate if they are not fully informed about the objectives of a project;

(c) Consultation. Decision-makers listen to the views of other stakeholders in order to improve project design prior to implementation or make necessary changes during implementation. The consultation process should involve the government authorities concerned, affected parties, donor agencies, mass awareness organizations and local and/or international organizations; and

(d) Participation. This aspect is an extension of consultation where directly affected groups become joint partners in the design and implementation of projects. They participate in "making" the decisions.

The lessons and experiences acquired by the water sector related to the creation of awareness and the implementation of the public participation policy are different. However, these activities are, generally speaking, new and without a given priority since each sector is obliged to concentrate on project implementation with the aim of meeting the urgent demands of society.

Initial Findings of Recent Efforts

Strategic planning places emphasis on the construction and improvement of scientific, efficient and systematic concepts and mechanisms. The initial findings resulting from recent efforts are detailed below.

a. Necessity to Rank the Human-centred and Nature-harmonized Concept as a Foundation

Due to guidelines provided by the Government, the whole of society has been made aware of the importance of human-centred involvement when making decisions. This is in line with the principle of benefiting the public instead of only considering economic or political interests as the leading target. The convenience and improvements have been obvious. What is more, the harmony between mankind and nature is also a critical indicator in the evaluation of water development project implementation. No stakeholders, even famous state-owned enterprises, can ignore their obligation to protect the environment. Local governments at different levels not only take the prevention of environmental pollution into consideration, but also regard it as a key motivation in promoting socio-economic development activities. At present, strategic planning highlights the reform of traditional measures in order to make environmental protection a profitable business, i.e., the more that the stakeholders invest, the more they can gain not only in terms of social benefits but also economic returns.

b. Importance of Eeinforcing the Systematic Mechanism

The systematic mechanism, strengthened by the strategic planning, mainly comprises the democratic consultation system, an efficient supervision system and an open evaluation system to allow different functions to be performed in making, implementing and assessing relevant decisions. Before making a decision, consultation workshops are organized at the community, district, provincial and central levels. Well-known researchers, institutions and universities, both at home and abroad, are encouraged to contribute a variety of methods and ideas. When implementing the projects, the supervision department is responsible for measuring the work output and results, discerning the drawbacks and advantages, and coordinating cooperation among the involved sectors. When implementing projects, evaluations are carried out in an open atmosphere to ensure acceptance of assessments, both from professional experts and relevant stakeholders.

c. The Integration of Water Resources Management

HRWRC experience of overall strategic planning and management suggests that the integration of water resources management into socio-economic development plans should be carried out at two levels: (a) the overall level, which refers to the links with national economic and social development; and (b) the sector level, which refers to integrated water resources management.

At the overall level, five elements are involved: (a) water planning coordination; (b) policy and registration development: (c) data and information management; (d) training; and (e) public awareness of integrated water resources management.

d. Additional Strategic Planning and Management Training

Strategic planning and management is crucial to integrated water resource management, and a great deal of successful experience has been accumulated in this filed. The representatives from the Haihe and Huaihe River basins regard strategic planning and management as a powerful tool in water resource management. However, since they have no experience and incomplete understanding of strategic planning and management, ESCAP has been requested to provide strategic planning and management training.

3

INTEGRATED WATER RESOURCES DEVELOPMENT FOR THE NADI RIVER BASIN

INTRODUCTION

Fiji is an island nation comprising more than 300 islands, of which 109 remain uninhabited. The two main islands (Vanua Levu and Viti Levu) support the majority of the total population of 775,000 with a sizeable percentage being in the urban centres of Suva (168,000), Lautoka (43,300), Labasa (24,100) and Nadi (30,884). The islands are predominantly volcanic and rise to an elevation of around 1,000 m with rivers and streams supporting the tropical rainforest on the windward side and extensively cultivated sugarcane farms on the leeward side.

All major economic activities including tourism are based on these islands. In contrast, the outer islands vary considerably in both geology and topography, ranging from smaller coralline islands to larger volcanic edifices that support smaller but significant populations.

The country's total land area of 18,272 km^2 is dispersed in territorial waters of about 141,800 km^2. The proportion of land to water is only 13 per cent, and even less when compared with the even larger Exclusive Economic Zone. Fiji enjoys two seasons (hot/wet and cool/dry) with rainfall averaging between 1,500 mm to 4,000 mm annually. The topographical effects mean that much of the rainfall falls within the windward side of the nation. Up to 80 per cent of the rainfall is recorded in the wet season and the remainder in the dry season.

Drought conditions are exacerbated during El-Nino episodes, leading to adverse impacts on all sectors of the economy. The development of water resources by the government are thwarted by a lack of clear and comprehensive legislation.

Fiji is blessed with an abundance of freshwater resources but the general population continues to experience inadequate water supplies for their daily chores or other needs such as agricultural and industrial uses.

Water resources development has not been given much attention despite experiences of extreme conditions, with half the year having above average rainfall and the second half suffering spells of drought. The Government's complacency towards providing a policy framework as a driving force in establishing a strategic approach to fresh water resources development and management has led to indiscriminate exploitation of these resources. The development and management of water resources is currently vested in a number government ministries and statuary authorities, each focusing on its own area of interest. Although some consultation between developers and the government agency responsible for legal approval, there is no authority or an apex body responsible for overall national development and management policies on water resources.

Of late, the Government's economic goals have been shifting increasingly towards encouragement of industrial activities, tourism and private sector promotion as a way of improving the standard of living. All these developments are dependent on the availability of quality water supplies.

The government's vision for sustainable development of this resource is: "To provide safe, adequate and affordable water and wastewater services for the total population of Fiji while equitably meeting the needs of other users (food production, energy generation and environmental maintenance) in a sustainable manner."

Water is a valuable natural resource that depends on the development of a strong political will that is related to highly sensitive land issues. However, there are many weaknesses in the development and management of the water sector that need to be addressed to ensure that socio-economic growth is not stalled. These include:

(a) The development of a Water Sector Policy

(b) Legislation enactment

(c) Institutional development

(d) Strong public sector coordination

(e) Development of human resources capabilities

The major challenge for Fiji is to maintain its freshwater resources from diminishing and to protect them from degradation.

Government Policy

The Government has embodied in its overall Strategic Development Plan in its commitment to the United Nations Millennium Declaration that was adopted by the United Nations General Assembly in 2000. The Declaration established a number of goals:

(a) Eradication of extreme poverty and hunger. The poverty level in the country is on the increase. In 1996, it was around 25 per cent, but recent figures indicate that it has increased to around 50 per cent. As a response to this situation, the Government recently increased its assistance to poverty alleviation including upgrading squatter settlements and other income-generating activities;

(b) Promotion of gender equality and empowerment of women. Women form more than half the population of the country. The Government recognizes the need to empower women in all phases of development and participation, both in the public and private sectors;

(c) Improvement of maternal health. Although life expectancy has increased from 67 years to 69 years, non-communicable diseases have been on the increase. The Government is committed to dedicating funds to health promotions and disease prevention. Waterborne diseases such as diarrhoea, dysentery, hepatitis and typhoid are due to the lack of a quality water supply;

(d) Ensuring environmental sustainability. Although Fiji is blessed with an abundance of natural resources (agriculture, forestry, fisheries and mining), it is interposed by climate extremes in the form of hurricanes, cyclones, flooding and drought. These extremes have serious economic, social and environmental consequences that require prudent macroeconomic management, proper land-use planning and water and watershed management. The Government's challenges are to address the degradation of land resources, the increasing risk of flood and the inundation of coastal areas, and unsustainable exploitation of land and water resources that undermines the quality of life; and

(e) Developing a global partnership for development. The Government is a member of the global village and recognizes the need for partnership in the development of its resources. It will therefore follow the indicators for achieving the Millennium Development Goal developed by various United Nation agencies, the International Monetary Fund and the Organization for Economic Cooperation and Development. The government is committed to achieving these goals and gives assurance that the policies in its strategic plan are consistent with the Millennium Development Goal. The proposed strategic plan for the Nadi River basin is aimed at contributing positively towards the Government's Strategic Development Plan and the Millennium Development Goals.

As part of the case study to look into the above aspects, the Nadi River basin has been selected as a model because:

(a) The Nadi River is vital to Fiji's economy as it continues to provide water for domestic uses as well as agricultural, hydropower and industrial production. Nadi town, which is located in the Nadi River basin, is the main commercial centre for the tourism industry. The importance of Nadi is both as the principal "gateway" to Fiji and as the base resort from where tourists travel to other parts of the country. The potential of Nadi Bay as a major resort centre and the development of Denarau Island as an integrated resort can form a model for the Tourism Development Area;

(b) Increasing population, rapid urbanization and industrial development continue to put high pressure on the water resources. Flood damage increase exponentially due to the rapid development taking place in the subject area;

(c) A strong strategic plan for the development and management of the Nadi River basin water resources and watershed management is required to ensure the sustainable and smooth operation of the present and planned structures;

(d) The hydrological functions within this watershed's forest and rangelands have been seriously disturbed by improper land-use practices. This has resulted in a number of onsite problems in the watershed areas such as the direct loss of productive agricultural land to gullies, landslides, surface erosion as well as the drying up of springs and wells. Downstream problems of flooding, drought and excessive siltation need to be addressed even more urgently; and

(e) As part of Nadi River basin plan, it has been proposed that the Nadi Watershed Management Project be established as a showcase project to strengthen the watershed management capability of the Land and Water Resource Management Department by implementing and constructing check dams in the tributaries of the Nadi River. It will also demonstrate the impacts of soil erosion and siltation within the catchment area, and this could be part of the department's education, training and research programmes for its personnel involved in the project.

Features of the Nadi River Basin

Hydro-geographical Characteristics

The subject area is covered with young volcanic rock as well as many types of older rocks scattered throughout the watershed area.

The monthly mean temperature fluctuates seasonally, the highest temperature occurring in February (27-30 C) and lowest in July (average 22 C). The relative humidity in this area ranges between 60 and 80 per cent. The seasonal variation of relative humidity is distinct as it is lower around September to November and higher in March and April.

Water Resources

As discussed above, the Nadi River basin has good water resource potential to support its socio-economic growth in the domestic, energy, agriculture and tourism sectors. There is one major river and a number of perennial creeks that service the Nadi watershed.

Political, Social and Economic Characteristics

The Government has the political will to ensure the sustainable development and management of the country's natural resources. It has the social obligation to ensure that the quality of life among its people is on a par with international standards.

The annual economic growth rate ranges from 5 per cent up to 8 per cent with political stability, and it is envisaged that this trend will continue for some time.

However, the Reserve Bank of Fiji at a recent financial summit in Nadi warned that Fiji's economic survival rested on agricultural development and revenue earned from agricultural export commodities and other natural resources. The expiry of leases for tenant farmers and the future of the sugar industry has been a recent cause of concern for the Government. The development of the Nadi watershed could therefore provide the required platform for the Government to achieve its objectives of economic growth, food security and poverty alleviation.

Water and Land Development

In the Nadi watershed, the difference between the current surface water potential and the water demand are close (surface water potential is 0.7 m^3/sec and water demand is 0.23 m^3/sec). However, with the rapid growth that is envisaged in this area, water demand by 2015 will be much higher than the potential.

The coastal plains of the Nadi watershed are currently under sugar cane cultivation.

However, these areas are already being considered for tourism development. A number of five star hotels are either being built or are on the drawing board awaiting approval from the authorization agencies. Water demand will continue to show an increasing trend. The total area of this watershed is 516 hectares and, with a population of more than 70,000 people, demand for land has already started to escalate. With development and job opportunities in this area, the migration of skilled people is expected to increase. The population explosion is expected to put pressure on land demand; thus, exploitation of land and water resources will be inevitable.

Environmental Issues Related to Water Resources

The Nadi watershed is highly vulnerable to natural and man-made disasters. Cyclones, flooding and severe drought have adversely affected the key sectors of the economy, particularly sugar cane production. During the past decade, cyclones and flooding have caused loss of life and damage amounting to millions of US dollars. The drought of 1997/1998 had a severe impact on the economy.

The main environmental problems within the subject area are land degradation, water pollution, climate change and sea level rise. The expansion of agricultural land is the principle cause of land degradation. To address these problems, the Government is finalizing an Environmental Management Bill that should provide the overall framework for identifying and addressing environmentally related issues. The Bill will provide policing of activities that bring about depletion of the natural environment.

Water-related Conflicts

The river system in the Nadi River basin runs across many different clans (matagali or landowners). Currently, there is sufficient water for all, and supplies from this watershed are transported outside the area for domestic and industrial purposes. Hydropower is also expected to be generated from within this watershed. However, with continuing population growth as well as developments that are being proposed, water demand will increase and water conflicts will arise.

Goals Linked to Social and Economic Development in the Context of National Development

A small, open economy and dependency on a few major exports makes Fiji's economy highly vulnerable to natural disasters as well as large swings in export prices. This has caused growth to be highly volatile. Fiji's economic growth has fluctuated erratically, with gross domestic product GDP varying from -5 per cent in 1985 to 3.8 per cent in 2001. Its real GDP per capita over the review period ranged from F$ 1,870.55 to F$ 2,590.64, the lowest rate being in 1987 and the highest in 1999. Agriculture remained the economic base for Fiji with heavy reliance on a single crop (sugar cane). A significant change in the structure of the economy during the 1990s coincided with a shift in economic policy from import substitution to export promotion.

The main challenge for the agriculture sector will come when the preferential price agreement for sugar ends in 2007. Sugar cane farming will no longer be viable unless the industry is reformed. The agricultural sector will need to diversify and look for other high-value crops for niche markets. However, such commodities will need a consistent water supply to ensure the production of good quality crops on a regular basis. The Nadi watershed, with its allowable water potential, can make an impact on export-oriented crops with outside markets at its doorstep. The tourism sector is booming with a number of five-star hotels either under construction or at the planning stage. This sector is supposed to be a potential major income earner for the country. With most of the hotels (including those still planned) located within the subject area, water demand will continue to increase.

Legal and Institutional Issues of Water Resource Management

The responsibility for water resource development and management is vested in different government agencies. However, very little coordination exists between those agencies at any stage of either the development or the management of water resources.

Each agency has its legislation in terms of management and empowerment to utilize water for its own purpose. Efforts made earlier to put in place legislation to develop and manage water resources under a unified law proved futile due to strong objections from some sectors with regard to the ownership of water.

A draft legislation entitled "Water and related land resources management" was proposed in 1976 as an umbrella Act to allow better coordination of policy and implementation. This was shelved because of a legal dispute between the Native Land Trust Board (custodian of native lands) and the State regarding recognition of riverbeds and banks in relation to their ownership.

Another attempt was made in 1987 by Professor Stanford D. Clark to review the draft Watershed Management Bill together with the draft Water and Land Related Resources Management legislation. This experienced a setback in terms of government priorities after the coups of 1987.

The Government also attempted to draw up a National Water Policy in 2002. Some work has already been done on this and it is currently with the office of the Chief Executive Officer of the Ministry of Lands and Mines.

The Environmental Management Bill (formally the Sustainable Development Bill) has been somewhat watered down to enable developers and managers to continue their exploitation of resources. This is currently being examined by the Parliamentary Working Group.

Strength and Weaknesses of the Current Legal and Institutional Framework

The current organizational setup within the Government is targeted towards national policies.

STRATEGIC PLAN GOALS

Vision for the Overall Integrated Water Resources Management Goal of the Nadi River Basin

The Fiji Vision Statement: "Nadi, a flourishing oasis and tourism capital of the Pacific by 2025" as the result of an integrated approach to management and development of the Nadi River basin. By 2025, the key infrastructure for effective management of the Nadi River basin water and related resources will be in place to support the initiatives to make it an environmentally sound river basin and the most important socio-economic centre of Fiji and the Pacific.

Integrated Water Resources Management Goal 1

This goal is to manage, develop and protect water and related resources to meet the needs of current and future generations.

a. Strategies to Achieve the Goal

The following areas need to be addressed:

(a) Policy and strategies for improving water management in the Nadi River basin;

(b) Programme of action for priority issues; and

(c) Policy and strategies to mobilize resources for an effective implementation of an IWRM programme in the Nadi River basin.

Stage 1: Preparations for medium-term development Stage 1 involves:

(a) Measures in 2004 to address the urgent need for flood control in priority areas of the basin, in order to enable tourism and other service sectors to attract private and foreign investment to the area. Carrying out a comprehensive flood mitigation programme to identify additional measures for the medium term (2005-2010);

(b) Measures in 2004 to ensure adequate water supplies to meet the urgent needs of expansion of tourism and other service sectors. Carrying out a comprehensive programme for the provision of water supplies for development and urbanization in the medium term (2005-2010);

(c) Establishing a detailed programme on natural disaster reduction and undertaking priority measures to improve preparedness and reduction, especially with respect to flooding and storm surges;

(d) Preparation of a detailed programme to promote eco-tourism and a report on corresponding impacts on water resources management in the basin;

(e) A priority programme to improve environmental conditions in the basin;

(f) A study on financial resources mobilization for the development of the basin with special attention to involving the private sector and international funding agencies;

(g) A study programme aimed at strengthening institutional capacity for water resources management at the Nadi River basin level and its linkage to national water resources policies; and

(h) A study on modernization of the irrigation system in the basin.

Stage 2: Preparations for long-term development Stage 2 involves:

(a) Establishing a Nadi River basin organization to facilitate coordination among stakeholders in order to enhance opportunities and efforts; and

(b) Undertaking measures in 2004 to address the urgent need for flood control in priority areas of the basin to enable tourism and other service sectors to attract private and foreign investment to the area. Carrying out a comprehensive flood mitigation programme to identify additional measures to be carried out in the medium term (2005-2010).

b. External Factors Affecting Goal Achievement

The most important factors affecting the progress of work in this area include:

(a) Willingness of all key stakeholders to participate and commitment by policy-makers to effect changes in economic structure and institutional set-up to meet emerging socio-economic demands;

(b) The lack of sufficient expertise in advanced technology and facilities for handling complex technical processes towards optimization of key stakeholders; and

(c) Continuity and consistency of leadership commitment, and availability of resources.

c. Cross-cutting Relationships

In order to ensure smooth cross-cutting relationships with all key stakeholders, regular consultations and the establishment of common targets will be necessary for evaluation purposes.

d. Performance Evaluation and Monitoring

It will be necessary to prioritize activities so that their results can be directly linked to service improvement according to customers' requirement and market conditions.

e. Performance Measures

The performance measures include:

(a) Policy and strategies to improve water supply and flood management in order to create an enabling environment for socio-economic development of the basin (June 2005);

(b) A report on a strategy to provide, manage and protection sources of water to meet the socio-economic goal of the Nadi River basin up to 2010 (December 2005);

(c) A programme of action for integrated water resources management in the Nadi River basin (December 2005).

Integrated Water Resources Management Goal 2

This goal is the operation, maintenance and rehabilitation of facilities safely, reliably and efficiently in order to protect the public investment.

Strategies to Achieve the Goal

The strategies include:

(a) Identifying and empowering the authorities with the necessary legislation, personnel, funds and other required resources for them to effectively undertake the operation, maintenance and rehabilitation of public investment in the basin;

(b) Developing and implementing a manual and procedures for reliable and efficient management of investment; and

(c) Encouraging the operating agencies within the basin to use the appropriate tools to manage their facilities effectively.

Integrated Water Resources Management Goal 3

This goal is the enhancement of organizational effectiveness. In order to achieve this goal, it will be necessary for all key agencies to improve the service provided to clients, particularly top decision-makers in the water sectors and the Government, and to strengthen the coordination among these agencies. For this purpose, it will be necessary to establish a coordination and information sharing mechanism among the agencies involved in the water sector; an institution should be assigned the responsible to provide such coordination.

In that connection, such an institution should assume its leadership role in coordination of water resources management, development and protection, and support an innovative growing workforce to respond to diverse needs. To improve business, it will be essential to involve decision-makers and key stakeholders in the formulation and implementation of this strategic plan as well as subsequent monitoring and evaluation.

a. Strategies for Achieving the Goal

The strategies include:

(a) Establishing an interim coordination mechanism and gradually developing it into an operational institution for coordination of Nadi River basin development;

(b) The formulation of a river basin plan for the development, use and management of the water resources;

(c) Development of capacity within the lead organization by engaging consultants and training locals in specialized areas;

(d) Encouraging stakeholder participation, from project formulation to the implementation and operational stages;

(e) Generating regular awareness among the stakeholders of the benefits of the project;

(f) Building pride of ownership and partnership in the project;

(g) Ensuring that adequate resources (physical/financial) are available at all times; and

(h) The provision of government support and political will.

b. Priority Activities

The priority activities include:

(a) Preparation of an Action Plan for two- and five-year periods. The Action Plan will be based on the urgent need for coordination of development activities in the Nadi River basin as well as priority water resource management issues at the national level. In the Action Plan, it will be necessary to emphasize the need to integrate a more balanced view of sectoral perspectives in the programmes of investment and operation of all water-related agencies. It will also be necessary to present the Action Plan to the Government for approval.

(b) Preparation and implementation of a capacity-building plan for key agencies, based on the requirements for achieving the Action Plan. The capacity-building plan will include building up a core expert group to support basin development. It is expected that external assistance will be provided for implementing all aspects of the capacity-building plan.

(c) Conduct a needs assessment for strengthening the legal and institutional framework for IWRM in the river basin. Prepare a coordinated training plan to address the priority training needs and assess the required training budget.

c. Performance Measures

The performance measures include:

(a) A capacity-building plan (December 2004); and

(b) Total number of trained staff - 50 with advanced technology expertise (December 2006).

IMPLEMENTATION, MONITORING AND EVALUATION

A monitoring and evaluation mechanism needs to be put in place to assist policy-makers, donor agencies and other stakeholders in periodically assessing the performance of the project.

Identification of Indicators and Benchmarks for Performance Measurements

a. Expected Time Frame and Necessary Resources

The expected time frame and necessary resources are:

(a) Finalize the pilot project within the Nadi River basin on the watershed management concept by the Department of Land and Water Resources Management in 2004 as a showcase project;

(b) Submission of a project proposal for 2005 Budget consideration by the central Government;

(c) Conduct a workshop to seek comments from the stakeholders and endorse the strategic plan in 2004;

(d) Establishment of a Working Committee in 2004 to ensure implementation of the strategic plan; and

(e) Seek government support for endorsement of this plan and funding for implementation.

b. Mechanisms for Reporting, Monitoring and Evaluation

Various mechanisms are in place for reporting, monitoring and evaluation purposes. The individual ministries and agencies involved in the development and management of water and associated resources are:

(a) Ministry of Agriculture, Sugar and Land Resettlement - responsible for development and management of water resources for agricultural use. Legislation for this purpose includes the Irrigation Act, Drainage Act, and Land Improvement and Conservation Act. There are no long-term water requirement plans in place;

(b) Ministry of Works - responsible primarily for the development and management of water for domestic and industrial purposes. Related legislation is the Water Supplies Act and Water Master Plan;

(c) Ministry of Health - responsible for ensuring good quality water for public health purposes. Legislation used is the Public Health Act;

(d) Ministry of Local Government, Housing, Squatter Settlement and Environment - responsible for the development and management of local government (Nadi Town Council) and the surrounding environment. Legislation used includes the Local Government Act, Environment Management Bill, and the Town and Country Planning Act;

(e) Ministry of Lands and Mines - responsible for regulating river and streams, and underground water. Legislation used is the Rivers and Streams Act. Currently, there is no legislation to regulate the development and management of underground water;

(f) Native Land Trust Board - responsible for the management of more than 90 per cent of the land in Fiji. Legislation used is the Native Lands Trust Act; and

(g) Fiji Electricity Authority - responsible for the provision of electricity supply, both rural and urban, through hydro-turbines or fuelled generators. Generation of power through windmills is currently at the initial stage. Development of hydropower within the Nadi River basin is also proposed for generating 2.8 MW of electricity.

Consultation Mechanisms and Initial Findings

All the stakeholders will be consulted at a workshop to outline the proposal contained in this strategic plan as well as the expected results and outcome.

a. Consultation Mechanisms in Place and Future Developments

As a way forward in ensuring that this becomes a living document, periodical meetings and workshops need to be scheduled to ensure that all stakeholders are kept up to date on the developments of each agency's work in the Nadi River basin. This will instill pride and a sense of ownership.

Moreover, the Government, other stakeholders and donor agencies will need to be attracted to ensure their support for this plan.

b. Initial Findings from Recent Efforts

Initial interviews with the major players involved in formulating this strategic plan have shown positive reactions. They have also indicated that the Department of Land and Water Resources Management can be the lead agency; however, more detailed consultations are needed with all the stakeholders in order to sell the idea and built good support. The development of ownership for the vision of this plan will have a strong bearing on its success.

The development of Fiji's National Water Policy has made some progress possible for the present Government's efforts to regulate the development and management of water resources. There is a need for support and assistance to ensure that this policy is adopted and allow the responsible agency to proceed with the formulation of legislation to regulate the development, utilization and management of water resources.

The stakeholders understand that water resources in the Nadi River basin can become a scarce commodity in view of current developments within the Nadi watershed, its rapid growth and the fact that Nadi could become the tourism capital of the Pacific.

3. Priority Action Programme

The workshop established a Nadi River Basin Working Team including the following core members: the Department of Land and Water Resources Management, the Nadi Town Council, the Tourism Bureau, NLTB and Land Department. It was agreed that the Working Team would aim at achieving the following priority outputs:

(a) Improved coordination of the different roles and functional activities such as awareness and disaster planning;

(b) Accelerated infrastructure development to attract investment;

(c) The establishment of supporting networks;

(d) The implementation of micro-catchment watershed management projects;

(e) Improved understanding among stakeholders through periodic consultations;

(f) Strengthened institutional framework of cooperation through the delegation of authority to related organizations;

(g) Improved information sharing and excellent coordination among all relevant stakeholders, including the development of an integrated database for the Nadi River basin (including GIS); and

(h) The establishment and implementation of effective risk management plans.

The Working Team agreed to carry out the following activities as part of the priority programme of action:

- Hold the first workshop to consider and approve the Terms of Reference of the Working Team before end of 2004;
- Formulate a detailed strategic plan for the Working Team on the basis of the current draft strategic plan for integrated water resources management in the Nadi River basin;
- Seek endorsement by stakeholders and communities;
- Implement, monitor and evaluate; and
- Report the findings to all relevant agencies, including the Government, for further support and resource mobilization.

CONCLUSION

The strategic planning and management process for the integration of water resources management in the Nadi River basin can contribute immensely to the socio-economic growth of the province and the nation. Although various modalities and methodologies are involved in developing a strategic plan, it appears that the people-centred approach can be very effective.

4

Integrated Water Resources Management in the Lao People's Democratic Republic

INTRODUCTION

The objectives set by the Government of the Lao People's Democratic Republic in its current Five-Year Plan (2000-2005) were to develop, modernize and industrialize as well as achieve a high quality of life for all. In the process of developing its strategic plan management (SPM), the Government specified three goals:

(a) Achieve an average annual economic growth of 8-8.5 per cent, which would translate into a per capita income exceeding US$ 500 by 2005;

(b) Devote at least 20 per cent of public expenditure to the social sectors, making education, health and social welfare the new "beacon sectors"; and

(c) Make poverty alleviation the major reference for socio-economic development by giving major attention to basic rural infrastructure and social services.

These goals clearly require support from the water sector, which comprises a number of agencies with resource development and management functions related to water and water resources.1 The sector is largely structured around the various uses of water such as hydropower, irrigation, water supply, fisheries and wastewater. These water subsectors developed their programmes to provide the support required for the Government's objectives.

The programmes, however, did not consider the potential for interaction between the policies and plans of subsectors. This was because no agency was given overall responsibility for managing water resources or the coordination of water resources development.

The Water Sector Strategic Plan has been developed to show how water sector activities support the Government's objectives of long-term socio-economic growth and to indicate the importance of these activities. The Water Sector Strategic Plan consists of a vision and mission for the water sector, a set of objectives and descriptions of key result areas necessary to achieve the objectives, with short-term goals and activities within those key result areas.

The Water Sector Strategic Plan is complemented by an Action Plan to address areas where current subsector strategies need supplementing in order to achieve sustainable and efficient use of the water resources. The strategy showed the area mostly overlooked was that concerned with cross-sector issues, which is the responsibility of the Water Resources Coordination Committee (WRCC). The Action Plan describes the required activities and shows their timing.

At meetings held to discuss the development of SPM, an issue of concern has been the difference between a master plan and a strategic plan. Therefore it is valuable at this point to state the difference between these two types of plans, as seen by the authors:

(a) Strategic plan. This is a plan with an underlying strategy or scheme for achieving an outcome, and which usually starts from the consideration of the goals, and institutional and policy issues. A strategic plan provides an outline structure within which a number of other more detailed development plans may be prepared;

(b) Master plan. This is a plan for controlling other plans at a lower level and usually does not involve institutional issues as a major item. A water resources master plan should follow the accepted strategy, and usually concentrates on issues related to the development of those resources.

National Water Sector Context

Some 80 per cent of the country lies within the Mekong River basin. The remaining 20 per cent drains through Viet Nam directly into the South China Sea. The major tributaries of the Mekong all have significant watersheds. In addition to the major tributaries of the Mekong, there are hundreds of small streams, most of which have a torrential regime during the monsoon season and very low or no flow during the dry season.

The total annual flow of water flow in the Lao People's Democratic Republic is estimated at 270,000 million m3, equivalent to 35 per cent of the average annual flow of the whole Mekong River basin. The monthly distribution of the flow of the rivers in the Lao People's Democratic Republic closely follows the pattern of rainfall: about 80 per cent during the monsoon season (May-October) and 20 per cent in the dry season (November to April). For some rivers in the central and southern parts of the country (particularly the Se Bang Fai, Se Bang Hieng and Se Done), the flow in the dry season is less, amounting to some 10 to 15 per cent of the annual flow.

The rivers outside the Mekong River basin - the Nam Ma, Nam Sam and Nam Neune - flow through Viet Nam into the South China Sea. The limited information on these rivers restricts assessment of their potential. The estimated inland water resources area totals approximately 723,500 ha, of which 200,000 ha are from Mekong River; 54,000 ha from other main rivers; 57,000 ha from reservoirs, 1,500 ha from swamps; 406,000 ha from rice fields; and 5,000 ha from fish ponds.

At the beginning of 2001, 140 gauging stations were in operation in the Lao People's Democratic Republic, nearly all of which are in the Mekong River basin. This amounts to an average density of about 1,350 km2 per station, the highest among the countries of the lower Mekong River basin. Records of flow for the Mekong River were started in 1904. However, the records for the Mekong tributaries have a much shorter history and vary considerably in length; most tributary stations are located on the lower reaches of the rivers where access is not too difficult.

Scope of the Study

The strategic plan management aims to maximize the benefits to the Lao People's Democratic Republic from the water resource management function. SPM comprises a number of component plans as specified in the draft policy:

(a) Principles of water and water resources management;

(b) Water source development and management;

(c) Public involvement;

(d) Financial resources for water source development and management;

(e) Water allocation, quality management and use;

(f) Data and information management; and

(g) Capacity-building and human resources development.

SPM will be based on a careful review of the effects of past development and will particularly take into account issues related to cultural minorities, including the stakeholders' preferred mode of development and implementation arrangements. The formulation of the development and management strategic plan will follow a consultative process to obtain the views of the stakeholders as well as the Government and its agencies.

Water Resources Potential and Challenges for the Water Resources Coordination Committee

The abundant water sources in the Lao People's Democratic Republic have the potential to support socio-economic development, especially the hydropower and irrigation subsectors. The hydropower potential of the country is large compared to other countries in the lower Mekong River basin, providing an opportunity to earn foreign revenue. The hydropower sector also has the ability to develop rapidly; it increased its production about five times from 247 million kWh in 1976 to 1,187 million kWh in 1999 when it exported 473 million kWh.

The Government has given high priority to investment in the irrigation subsector as agriculture forms the foundation of national economic development, and is necessary for food stabilization. In addition, some 85 per cent of the population lives in rural areas. From 1976 to 2000, the area of irrigated dry season rice increased about 40 times from 2,700 ha to 110,000 ha.

The irrigation subsector also significantly increased the average yield of rainfed paddy rice from 1.43 mt/ha in 1976 to 3.27 mt/ha in 2000. However, water resources development is still at a low level: the irrigated area is only 20 per cent of the national paddy area while hydropower production is still at only 2 per cent of its potential of 30,000 megawatts.

Development in other sectors is still at a low level compared with hydropower and irrigation. In addition, several problems are faced by the water sector. These include:

(a) Unusual rainfall patterns in some years, plus high evaporation, flooding and drought in some of the main agricultural areas of the country;
(b) The impact of shifting cultivation on water resources, although this activity has been significantly reduced; and
(c) Conflicts of interest for management within the sector since most water subsectors are still responsible for multiple roles of regulator, manager and service provider.

As the Lao People's Democratic Republic is a least-developed country with rich water resources, the most important challenges for WRCC in carrying out its coordinating role include:

(a) Strengthening the legal framework for effective and harmonious integration of water resources management, development and protection activities into the socio-economic development process of the country, in particular to meet national priorities;
(b) Enhancing and consolidating the existing systems and foundation to operate, maintain and rehabilitate facilities safely, reliably and efficiently in order to protect the investment for public benefit; and
(c) Prioritizing the capacity-building needs in order to enhance the organizational capacity and effectiveness of the water resources coordination system.

SUMMARY OF THE GOALS LINKED TO SOCIO-ECONOMIC DEVELOPMENT TARGETS

Overall Socio-economic Development Context of the Country

Within the water sector, the irrigation and hydropower subsectors remain the most important in terms of investment. Dry-season irrigation, especially small pumping irrigation schemes, has been given the highest priority in the short term because rice production was seriously affected by droughts and flooding during several consecutive years. Pumping schemes have successfully increased income in rural areas while delegating management to local authorities. The ongoing programme of transferring irrigation management to the beneficiaries has become critical to the sustainability of these schemes. This transfer of responsibility will need to be intensified in view of the number of pumping schemes being implemented.

The development of medium and large-scale hydropower schemes for energy export was affected by the economic crisis in Thailand, which has been the major market to date. In the short term, hydropower development will focus on medium, small and micro schemes for remote urban centres and rural areas.2 In the water supply and sanitation subsector, the target is to provide basic facilities to all provincial capitals and access to clean water for 70 per cent of the rural areas by the end of the century.

In 1997, real GDP growth was estimated to be 7 per cent, slightly higher than in 1996. This result was mainly due to the contribution by agriculture, which experienced favourable weather conditions after severe and extensive flooding in 1995-1996.

As the economic structure of the Lao People's Democratic Republic is dominated by subsistence agriculture, the effect of the current regional economic situation is not expected to be as severe as in some other countries. The subsistence economy of most rural areas will be less affected than the urban centres. Sectors most affected by the regional economic crisis will be foreign direct investment and timber exports due to the slowdown of construction activities in the region, particularly Thailand. Depreciation of the national currency, the kip, will seriously affect the national budget in the short term. The gap between the resources available and the socio-economic development requirements will widen. The Government's ability to close this gap is also influenced by the high fertility rate of 5-6 per cent (compared with the population growth rate of 2-3 per cent), and the high incidence of poverty.

Three essential elements are embodied in the national policy framework:

(1) Growth with equity while safeguarding the social, cultural and political identity of the country;

(2) National integration; and

(3) Shifting from a landlocked to a land-linked country in order to enhance economic and trade cooperation of the Greater Mekong Subregion.

The long-term development directive for 2000-2020, adopted by the sixth Party Congress, is to permanently rise above least developed country3 status. To achieve this result, the Government has set three overall goals for the five-year period of 2000-2005. The eight priority programmes of the Government stemming from these goals are at various stages of formulation by line ministries. These programmes are:

(a) Food production
(b) Stabilization of shifting cultivation
(c) Commodity production
(d) Infrastructure development
(e) Rural development
(f) Human resources development
(g) Services development
(h) External economic relations

In the water sector, irrigation contributes to self-sufficiency in food, agricultural production and a reduction in shifting cultivation. Hydropower contributes through export power projects and by reducing migration to urban centres through increased availability of electricity, rural development and income redistribution. Navigation improves transportation links, particularly with China, Myanmar and Thailand. Fisheries are a major source of food and contribute to foreign earnings. Urban water supply meets industry and urban population needs. Rural water supply and sanitation improvements increase the health and living standards of rural communities and assist tourism. Drainage, solid waste and sewerage provide for industry and development in urban areas. Water also maintains the natural environment and hence the welfare of the Lao people by providing fish and animal protein as well as aquatic plants. Wetlands are a refuge for fish, animals and birds.

Article 19 of the Constitution of the Lao People's Democratic Republic is specifically devoted to the protection of the environment. Action to protect the environment is considered a duty of all organizations and citizens. Recognizing the strategic value of natural resources in the socio-economic development of the country, especially forests, land and watei resources, the national environment policy places primary focus on the preservation of these resources while maximizing revenue generation and maintaining the livelihood of rural population.

Opportunities and Threats for Water Resources Development from the Perspective of the Water Resources Coordination Committee

The Lao People's Democratic Republic has plenty of surface water available but its quality is not always good. Land-use changes appear to be altering the pattern of stream flows (stream flow regime) and some supply shortages are occurring. Nevertheless, in general there is currently little conflict over water use. Increased water storage will be needed to support development. This additional storage will need careful management to prevent unwanted effects.

Throughout the Lao People's Democratic Republic, sediment is the primary river pollutant, although close to the major developed areas nutrients are also an issue. Some nutrients and some sediment are needed to support the aquatic environment; however, too great a quantity of nutrient may lead to algal blooms and encourage unwanted weed growth. Land that has little or no vegetation is the main source of sediment. The issue of sediment sources is being addressed in upland areas through watershed management activities.

Current water quality is mostly good, although the trend is towards increasing pollutant loads. Because of this, plans for increasing agricultural production through the application of high phosphate fertilizers should be implemented with caution.

Increasing population and higher population densities will increase health risks unless accompanied by good water management. Disease risks include dysentery, dengue fever, malaria, cholera and typhoid.

There is little information on the extent and capacity of groundwater systems. However, some use is already being made of the systems for water supply. This resource could be significant in some areas, particularly those underlain by limestone where large extraction rates could be expected from a single correctly placed bore. Such areas with high groundwater flow rates will also need careful management to avoid pollution from inappropriate land use.

Flooding is a major issue, with damage occurring in both urban and rural areas. Floods cause communal disruption, economic losses and increase health risks. On the other hand, floods bring some benefits such as building rich floodplains and replenishing soil moisture. Flood plain management could be considered as a way to reduce flood damage by guiding sensitive development away from flood-prone areas.

The population needs fish for protein. Fisheries also employ significant numbers of people and contribute economically. Fish populations are declining, and this is becoming a serious concern. The reasons for this decline are not yet known. Changes in land use, declining water quality, construction of dams and other barriers to migration, and high levels of capture are all likely to be involved. Reservoir fisheries are being developed to increase production.

Vision Statement of the Water Sector

A Vision Statement is a phrase that expresses the long-term objective and goal of the water sector. The Vision Statement suggested by working group members for the sector is "coordinated development of water and water resources for health, wealth and happiness". Some of the reasons and thinking behind this statement are worth elaborating. Coordinated development refers to the need to consider the interests and aspirations of all stakeholders in water and water resources. Water and water resources has a specific meaning in the Water and Water Resources Law, and therefore refers directly to that law. Health, wealth and happiness refer to the desired outcomes from development for the population of the Lao People's Democratic Republic. These three words imply sustainable development, or a balance in economic, social and environmental outcomes. The phrase also has some commonality with the national motto: "Peace, independence, democracy, unity and prosperity"

LEGAL AND INSTITUTIONAL FRAMEWORK

Legal and Institutional Context

Since 1991, the Constitution has become the foundation document for the socio-economic management of the country. It forms the basis for the functioning of a multi-sector economy, and provides the conditions for the education, culture and health of the Lao people.

In June 1998, a National Commission for Constitution Review was established to assess the impact of the new Constitution and to propose possible amendments in order to adapt to the changing domestic, regional and global environments, and bring the Constitution in line with the new economic mechanism. The draft of the amended Constitution underwent a nationwide public consultation process to ensure that the amendments met the real needs of the Lao people in the current economic period.

The Water and Water Resources Law of 1996 sets out a legal framework for development in the water sector. Many issues in the law, particularly the roles and responsibilities of various agencies for specific activities such as water allocation and the process for licensing water users, need to be developed. There is an urgent need for the development of further legislation or decrees for subsectoral activities, as well as the necessary legal documents to accompany the law and make it effective. Support given to the Government in this area should be undertaken in a counterpart relationship with the Department of Legislation within the Ministry of Justice, as this agency has the final responsibility for issuing decrees.

External assistance is also required by water subsector agencies in drafting regulations. This assistance should cover both drafting and the operational and enforcement aspects of regulations. A pre-requisite for success in this field is capacity-building throughout the water sector because this is a new activity and because the water sector traditionally focuses on development rather than management and regulation.

Current Situation and Perspectives of the Legal and Institutional Framework

The WRCC Secretariat is the executing body for WRCC and comprises a director and a number of technical specialists. It is responsible for any task assigned by the WRCC Chairman. It is also responsible for administrative and financial matters and for the management of WRCC assets. WRCC is composed of the Vice-President of the Science, Technology and Environment Agency as chairman together with representatives of the following organizations nominated by the Prime Minister's Office:

(a) Ministry of Agriculture and Forestry (Vice-Chairman)
(b) Ministry of Industry and Handicrafts (member)
(c) Ministry of Communications, Transport, Post and Construction (member)
(d) Ministry of Public Health (member)
(e) Lao National Mekong Committee (member)
(f) Science, Technology and Environment Agency (member)

Due to the necessity for wider coordination as well as contributions from water users in all aspects, additional representation from the Ministry of Justice, Lao Women Union and Lao Front for National Construction to the WRCC was needed. Under the Prime Minister's Office Notice No. 1908/PMO, dated

29 October 1999, a representative from the Lao Women's Union was assigned to join the Committee.

The two other assignments are still in the consideration process. At the same time, an increase in secretariat staff is needed to expand both quality and quantity of work. The development of sound legislation, regulations and guidelines is one of the primary means by which WRCC can play its coordinating role within the water resources sector and within this project. The process must be open and consultative, involving stakeholders at the central, provincial and local levels. WRCC already has some experience in this respect but further capacity-building for policy analysis and development is needed. WRCC should establish an inter-ministry working group for legal and institution development. It should seek the comments of stakeholders through the awareness and consultation process described below. Legal and institutional changes need to be accompanied by detailed implementation plans that indicate agency responsibilities and, where necessary, provide for further capacity-building to allow implementation to be successfully carried out.

Strengths and Weaknesses of the Current Legal and Institutional Framework

Present organizational arrangements at the national and provincial levels generally support the achievement of national policies. However, some agencies have conflicting roles such as those of regulator and service provider. This could result in ineffective implementation of government policies as well as enforcement of rules and regulations.

The Government has a policy of decentralizing planning and development to the provinces, with broad control and guidelines set at the central level. This is to ensure that local conditions and needs are appropriately considered. In the case of irrigation, for example, the head of the Provincial Irrigation Service is responsible to the head of the Provincial Agriculture and Forestry Service Office who, in turn, is responsible to the provincial governor. The intervention of the Department of Irrigation in the provinces is related mainly to reviewing the annual irrigation development plans and budgets proposed by the provinces, providing technical assistance to the Provincial Irrigation Service, and monitoring the implementation of the annual work programmes. There is no involvement of a regulator to consider whether these plans conflict with natural resource values or to set operating conditions.

Although devolving the responsibility from the central Government to the provincial authorities is helpful in gaining greater community input, the task of achieving broad environmental outcomes and maintaining technical performance is made harder. The control of natural resources and cross-sector coordination needs to be considered in planning this decentralization. The current institutional problems in the water sector are mainly related to the lack of coordination between agencies within the sector and with those of other sectors, and to the loose lines of communication and coordination between the national agencies and their provincial counterparts. WRCC, as a national apex body, is mainly aimed at improving the coordination of multi-sectoral activities that involve various water uses, and also at defining and managing water allocations.

This is an important initiative for coordination at the national level as it enables the Government to overcome the current fragmented management of water resources. The tasks of WRCC are greater than just coordinating the activities of the subsectors. WRCC should be vested with the responsibility of allocating water resources among the various water users and the environment. To undertake this significant task in addition to coordination, WRCC will require a strong, dedicated and experienced secretariat. This will only occur with sustained support from donors such as the Asian Development Bank and a commitment from the top levels in the Government. Measures required for coordination at the provincial level should not be viewed as being of lower importance, as the greater part of the national budget (covering project planning and execution) is managed by the provincial administrative authorities. The first step would be to assess the present institutional set-up at the ministerial and provincial levels, and to conduct a diagnostic study of each targeted river basin. The findings of these studies could be used as a basis for designing a river basin management body tailored the development mix in each basin. The policy of decentralization is now being implemented. Decision-making authority is being moved from central government to the provincial level. This includes formulating annual work programmes, project planning, implementation, and operation and maintenance. Service provision is also being transferred to local agencies and communities, such as provincial water supply agencies, irrigation water user associations and village water supply management committees. State-owned service providers are vested with greater operational and financial autonomy but are still limited in matters related to personnel management, salaries and incentives policy. State-owned service providers are usually operating within the administrative boundary of the province.

As a result, development objectives and investment performance depend on local commitment and capacity. The need for institutional strengthening and capacity-building is increasing as a result of the changes now underway. Efforts in this area should, therefore, focus on the provincial and local levels, while incentives for government employees should be biased in favour of the provinces. In all aspects of capacity-building at the central, provincial and local levels, efforts should consider and promote the capabilities of women. The aim should be to correct the imbalance in women's involvement in planning, management and decision-making concerning water resources development and management. The specific skills of women in the essential human aspects of this work (e.g., coordination and conflict resolution) should be encouraged.

Mission and Vision Statements of the Water Resources Coordination Committee

The Mission Statement suggested for the water sector is "Protect, manage and develop water and water resources to achieve improved quality of life in the Lao People's Democratic Republic". As with the Vision Statement, the words used have specific meanings. Protect refers to the need to ensure the continuing supply of high-quality resources. Manage refers to the activities of managing the resources, including managing access to the resources, managing developments that use the resources and managing the institutions with responsibilities related to the resources. Develop refers to converting the resources to new uses and to a mature state for the purpose of improved quality of life. The 1999 WRCC Mandate defines the rights and duties of WRCC. That list of rights and duties, although broad, is mainly directed towards actions such as "study, monitor, coordinate and advise." It also clearly indicates an important role for WRCC in all the integrated water and resources management functions mentioned below.

The 2001 Decree to Implement the Law on Water and Water Resources, issued by the Prime Minister, defines the structure of water resources planning and management at the national and river basin levels. The Decree states that WRCC is: "Responsible for coordinating line agencies in drafting of strategies and action plans, programmes and regulations necessary for the planning, management, use and protection of water and water resources. It is also responsible for monitoring, control, promotion and reporting on the implementation of activities related to water and water resources." In this connection, the WRCC Vision Statement is "to attain the modern status of a State coordination agency to provide advisory services to the Government in decision-making, based on a more balanced treatment of sectoral water perspectives, and to mobilize the public in their participation in the development, management and protection of the water and water resources to achieve improved quality of life in the Lao People's Democratic Republic".

STRATEGIC PLAN GOALS

Overall WRCC Goal

The overall WRCC goal is "towards consensus building and sustainable development of water and water resources in the country, particularly for the priority areas identified by the Government". Within the context of this overall goal, WRCC will attempt to achieve the following three main mission goals:

WRCC Mission Goal 1: Improve and modernize the legal and technical framework for managing, developing and protecting water and related resources to meet the needs of current and future generations.

WRCC Mission Goal 2: Enhance and consolidate the existing systems and foundation for operating, maintaining and rehabilitating facilities safely, reliably and efficiently, in order to protect the public investment.

WRCC Mission Goal 3: Enhance the organizational capacity and effectiveness of the water resources coordination system.

The three mission goals are linked to the three above-mentioned priorities of the country, in terms of ensure high economic growth, poverty alleviation and strong framework for sustainable development.

WRCC Mission Goal 1

WRCC mission goal 1 is to "improve and modernize the legal and technical framework for managing, developing and protecting water and related resources to meet the needs of current and future generations". To achieve this goal, WRCC will work in cooperation with all key agencies and stakeholders towards effectively managing, developing and protecting water and related resources for agricultural, municipal, industrial, rural, hydropower, recreational, and fish and wildlife purposes. WRCC seeks to gain the greatest benefit from existing resources in a manner that recognizes competing interests, uses sound conservation practices and is efficient. Within the context of the Water Resources Law and existing legal framework, WRCC will pursue its coordinating roles aimed at improving water and related resource management, development and protection activities. The framework for the coordinating roles includes:

(a) Administering the implementation of the Water Resources Law as well as the development of laws and regulations to ensure balanced treatment of sectoral perspectives, optimum water resources development benefits and an effective decision-making process;

(b) Formulating a policy and strategies to improve the water resources planning system of the country, with emphasis on the river basin water resources planning system;

(c) Formulating a policy and strategies to improving the efficiency of water use; and

(d) The promotion of integrating applied sciences and technology into water resources management.

a. Improving the Implementation of the Law on Water and Water Resources

A review of current water resources policy and legislation was initiated in 2004 in order to recommend priority topics for further policy and legislation development and, with WRCC approval, assist WRCC to prepare a draft policy, legislation, regulations, guidelines and other documents for submission to the Government for approval in 2005.

Strategies for Achieving the Goal

To assist in this review process, arrangements for effective inter-ministry and key stakeholder involvement will be made to establish an appropriate mechanism, such as a Policy and Legislation Working Group. In this process, all key ministries, agencies and stakeholders will be requested to prepare a brief plan (2-4 pages) listing and prioritizing policy and legal issues to be addressed, opportunities for more effective management, development and protection of water and water resources, and, if possible, to present strategies for dealing with the critical issues. For each significant issue or opportunity, the plans should briefly address the following aspects:

(a) The history of the issues, including the customers and key stakeholders involved;

(b) Current status of the issues;

(c) Recommended strategies, actions and options for each agency or stakeholder, including goals to measure progress;

(d) Implications of the proposed actions and intended results; and

(e) External factors (social, policy, legal etc.) affecting achievement of the strategy and progress in addressing the issue.

On the basis of inputs from key ministries, agencies and stakeholders, WRCC in cooperation with the Policy and Legislation

Working Group will analyse and synthesize the issues into a framework for the development of policy, legislation, regulations, guidelines and other documents for submission to WRCC and the Government for approval.

External Factors Affecting Goal Achievement

The most important factors affecting the progress of work in this area include (a) willingness of all key stakeholders to participate, (b) the lack of expertise in the policy and legislation areas of key stakeholders, including the limited staff of WRCCS, (c) the understanding among expatriate experts of local sensitivity in the implementation of policies and enforcement of legislations, and (d) public awareness. On the other hand, given the limited resources available to the country, and particularly to all key agencies, it will be difficult to ensure active participation by all key stakeholders.

Cross-cutting Relationships

To ensure smooth cross-cutting relationships with all key stakeholders, regular consultations and the establishment of common targets will be necessary for evaluation purposes.

Performance Evaluation and Monitoring

It is necessary to prioritize activities so that their results can be linked to the priority areas of socio-economic development of the country, and in particular stable and sustainable economic growth and poverty eradication goals. For example, specific target dates with appropriate indicators should be identified for priority areas of economic development and poverty eradication, particularly in those areas facing water constraints.

Performance Measures

The performance measures include:

(a) A comprehensive review of priority issues in water resources policy and regulations in the Lao People's Democratic Republic;

(b) Formulation of a strategy to develop policy, legislation, regulations, guidelines and other legal documents for submission to the Government for approval; and

(c) The start of operations (expected in 2005) by the Inter-Ministerial Working Group on Policy and Legislation Improvement.

b. Improving the Water Resources Planning System

This aspect requires careful coordination of the river basin water resource planning system, including detailed procedures and roles for WRCC, river basin committees and other key participants in the preparation, approval and implementation of river basin plans. This plan is expected to serve as the guideline for provinces, local authorities and other central agencies in conducting detailed project planning in a coordinated manner. The plan will also specify agency responsibilities, including that of monitoring plan implementation.

Strategies for Achieving the Goal

In view of the limited resources available for modernizing the water resources planning system in the Lao People's Democratic Republic, it is necessary to devise a number of strategies to respond to the priority needs of socio-economic development, particularly stable and sustainable economic growth and poverty eradication goals. This can be achieved through the following three-pronged approach:

(a) A pilot river basin study. The Nam Ngum River basin has been selected as the river basin for the pilot study;

(b) Pilot community-based water resources development areas for poverty eradication. These areas could be within the Nam Ngum River basin or in other areas, depending on the critical dimension of water resources constraints; and

(c) A programme to expand basin planning systems to other river basins and poverty eradication areas. Depending on availability of resources, this programme could have a number of target dates, but the overall framework will be governed by the long-term development goals of the country, such as Vision 2020.

External Factors Affecting Goal Achievement

The most important factors affecting the progress of work in this area include (a) the willingness of all key stakeholders to participate, (b) commitment by the Government and key funding agencies, (c) the effectiveness of technology and know-how transfer to local staff, particularly WRCC technical staff, and local consultancy firms, and (d) the "brain drain".

Cross-cutting Relationships

In order to ensure smooth cross-cutting relationships with all key stakeholders and the effective transfer of technology and know-how, regular consultations and establishment of common targets will be necessary for evaluation.

Performance Evaluation and Monitoring

While detailed implementation of the Nam Ngum River basin plan has been established with clear target dates of delivery, no plan has been established for the second and third prongs of the above strategies. In addition, indicators should be established to ensure sustainability of the new water resources planning system.

Performance Measures

The performance measures should include:

- A pilot river basin study of the Nam Ngum River basin;
- A report on the development of a river basin water resource planning system, including detailed procedures and roles for WRCC, the Hydro-Power Organization, river basin committees and other key participants in the preparation of river basin plans, including plan approval and implementation procedures;
- An integrated management plan for the Nam Ngum River basin. This plan is expected to serve as the guideline for provinces, local authorities and other central agencies in conducting detailed project planning in a coordinated manner. The plan will also indicate agency responsibilities, including the responsibility for monitoring plan implementation;
- Pilot community-based water resources development areas for poverty eradication.
- A draft report on detailed procedures and roles for related agencies and stakeholders in the planning of community-

based water resources development, including plan approval and implementation procedures;

- A report of community-based water resources development for pilot areas for funding and implementation;
- A programme for expanding basin planning systems to other river basins and poverty eradication areas; and
- A strategy for applying the planning systems to other river basins and other priority areas.

c. Formulation of Policy, Strategies and Plans to Improve Water-use Efficiency

WRCC will assist related agencies in the planning and development of water and power projects that improve the efficient use of limited water supplies in order to meet the socio-economic development needs.

Strategies to Achieve the Goal

WRCC will aim at addressing the following three areas:

- A policy and strategies to improve water supply and sanitation in order to achieve the national goal in water supply and sanitation
- A policy and strategies to improve water management in the Nam Ngum River basin
- A policy and strategies to replicate the experiences for other areas of the country With regard to the Nam Ngum River basin, activities will include:
- Recalibrate the river basin model for short-term (weekly and daily) operational planning of the Nam Ngum 1 Reservoir for hydropower generation and flood mitigation during the monsoon season as part of an integrated flood model, consisting of:
  (a) the Nam Lik River down to the confluence with the Nam Ngum River;
  (b) the Nam Ngum 1 Reservoir and the upstream catchments area;
  (c) a hydro-dynamic model of the area downstream of the Nam Lik/Nam Ngum confluence; and

(d) an economic model based on flood damage cost functions and hydropower benefits; and

Application of the integrated flood model to:

(a) define the optimum flood control operation rules for different flood scenarios in the Nam Ngum River basin and the Mekong (backwater) as general guidelines;

(b) demonstrate the hydrologic/hydraulic inter-relationship causing flooding in the Vientiane Plain; and

(c) propose other flood mitigation measures to alleviate frequent flood damage.

External Factors Affecting Goal Achievement

The most important factors affecting the progress of work in this area include:

(a) willingness of all key stakeholders to participate, particularly the public, in demand management;

(b) the lack of sufficient expertise in advance technology and facilities on handling the complex technical process towards optimization of key stakeholders, including the limited staff of WRCC;

(c) the willingness of key expatriate experts to transfer technical know-how; and

(d) continuity and consistency of leadership commitment and availability of resources.

Cross-cutting Relationships

In order to ensure smooth cross-cutting relationships with all key stakeholders, regular consultations and the establishment of common targets will be necessary for evaluation.

Performance Evaluation and Monitoring

It is necessary to prioritize activities so that their results can be directly linked to service improvement according to customers' requirement and market conditions.

Performance Measures

(i) Policy and strategies to improve water supply and sanitation so as to achieve the national goal in water supply and sanitation Submission of a strategy to provide, manage and protect sources of water to meet the national water supply goal in 2020.

(ii) Policy and strategies to improve water management in the Nam Ngum River basin Submission of a report on sharing water resources in the Nam Ngum River basin, and the preparation of a detailed plan for flood management and preparedness in the Nam Ngum River basin.

(iii) Policy and strategies to replicate the experiences for other areas of the country Proposed policy and strategies to improve efficiency of water use in other areas of the country.

d. Promoting Integration of Applied Sciences and Technology into Water Resources Management

Application of sciences and technology to water resources management will facilitate decision-making, improve transparency to ensure better coordination, and enhance accountability.

Strategies to Achieve the Goal

Apart from improvement of the planning system, the flow of information is important. The strategies in this area will include:

(a) Improving the flow of information for decision-making, and enhancing the effective and efficient use of resources for basic data collection;

(b) Coordinating the development of a detailed data and information management strategy. This includes an assessment of current practices and issues in the use of technology in decision-making and recommendations for introducing greater coordination and accessibility of water resources data. Review, update and disseminate the metadata directory established under previous technical assistance. Prepare and maintain a water resources information system for WRCC; and

(c) Reviewing the need for additional meteorological and hydrological data, within the available budget, in order to enhance the quantity and quality of data for calibration and future applications of the model, and recommending site locations, purchases and supervision of the installation of new equipment.

External Factors Affecting Goal Achievement

The most important factors affecting the progress of work in this area include:

(a) Willingness of all key agencies to participate in the establishment of coordinated data systems, which should not take away the incentives and ownership of all data collection agencies;

(b) The lack of effective mechanisms for the transfer of data and information;

(c) The lack of sufficient expertise and guidelines to link advanced technology to decision-making, especially in the case of the limited WRCC staff;

(d) Willingness of key expatriate experts to transfer technical know-how; and

(e) The lack of regular communication between decision-makers and technical staff, and consequently the lack of continuity and consistency in leadership commitment and availability of resources.

Cross-cutting Relationships

To ensure smooth cross-cutting relationships with all key agencies, regular consultations and the establishment of common targets will be necessary for pilot projects on information management improvement and their subsequent expansion.

Performance Evaluation and Monitoring

It is necessary to prioritize activities with target dates and indicators so that their results can be directly linked to the expectations of decision-makers in terms of policy and strategy implementation in water resources management.

Performance Measures

(i) Strategy on data and information management for decision-making in the water resources sector. Preparation of a report on a comprehensive review of priority issues in water resources policy and regulations in the Lao People's Democratic Republic.

(ii) Strategy for effective collection of basic information and data on river basins Preparation of a strategy for data collection and management in the Nam Ngum River basin and possible application to other river basins.

WRCC MISSION GOAL 2

WRCC mission goal 2 is to "enhance and consolidate the existing systems and foundation for operating, maintaining and rehabilitating facilities safely, reliably and efficiently, in order to protect the public investment". To achieve this goal, WRCC will focus on operating, maintaining, and rehabilitating existing water resources facilities so that they continue to provide project benefits. These facilities provide power and water supply delivery systems that serve agricultural and municipal requirements, in addition to having benefits for recreation, fish and wildlife as well as flood control. By ensuring that facilities are safe, cost-effective and reliable, WRCC can assist the agencies concerned to provide the best possible project benefits while protecting public health, sustaining environmental values, and providing timely and economical services to customers and thus protect the public investment.

WRCC will assist the related agencies to conduct oversight reviews and planning to ensure timely operation and management procedures, including control and communications systems for continued safe and reliable operation of existing facilities. In this connection, it is expected that the related agencies will be able to verify reliability, improve maintenance practices, and identify and implement risk reduction actions.

a. Ensure Effective Operation of Facilities

It is important that related agencies comprehensively review, update and implement standard operating procedures and related operating guidance to ensure that existing facilities provide project benefits in a safe and reliable manner.

Strategies for Achieving the Goal

WRCC will, through the implementation of the related laws and development of regulations, encourage all operating agencies to use appropriate tools, technologies, innovations and research to operate facilities effectively, including internal auditing systems.

External Factors Affecting Goal Achievement

Natural events such as flooding, drought, earthquakes and fires have a serious impact on the ability of the related agencies to operate facilities and provide project benefits in a safe and reliable manner. Funding limitations may restrict accomplishments.

Cross-cutting Relationships

In accomplishing this long-term goal, WRCC will have to ensure that appropriate procedures and investments are allocated the necessary resources to undertake these activities and encourage the participation of stakeholders.

Programme Evaluation and Performance Improvements

The establishment of indicators and monitoring of incidents will be necessary for ensuring effective evaluation of performance improvements.

Performance Measures

Reporting on emerging issues in the operations of existing facilities in the water sector. Proposed strategy for improving the efficiency and effectiveness of existing facilities.

b. Effective Participation by the Public

Participation by the public is essential to enhancing the benefits of existing facilities and minimizing risks and impacts of disasters. In addition, effective public participation will result in better investment strategies in terms of resources mobilization and investment scheduling.

Strategies for Achieving the Goal

The following approach is necessary:

(a) Develop an awareness and consultation strategy. This strategy will

(i) identify target groups and their information needs as well as an appropriate and cost-effective means of providing information and creating awareness,

(ii) identify modalities and an action plan for consultation with stakeholders regarding draft policy and legislation, and other important consultation topics, and

(iii) identify awareness and consultation roles of central government agencies and the River Basin Committee in relation to all major target groups.

(b) Carry out awareness and consultation responsibilities in accordance with the agreed strategy.

(c) Play an awareness and consultation role at the river basin level.

(d) Develop and implement national guidelines for dam safety.

Cross-cutting Relationships

As the overall coordinating body in water resources management, WRCC will assist other agencies in the development of the above strategies and guidelines, including the establishment of an appropriate mechanism for implementing the strategies and guidelines.

Programme Evaluation and Performance Improvements

Detailed indicators will be developed to monitor the implementation of the related strategies and guidelines.

Performance Measures

The performance measures should include (a) the establishment of working groups on public awareness and participation, and (b) a plan and strategy to enhance public awareness and participation.

WRCC Mission Goal 3

WRCC mission goal 3 is to "enhance the organizational capacity and effectiveness of the water resources coordination system". WRCC needs to be strengthened to become an effective organization in order to achieve the mission goals. Under this mission, WRCC will improve the service provided to clients, particularly top decision-makers in the water sector and the Government, strengthen its leadership role in the coordination of water resources management, development and protection, and support an innovative and growing workforce to respond to diverse needs. To improve its operation, involvement of decision-makers and employees of WRCCS in the formulation and implementation of the strategic plan and subsequent monitoring and evaluation is essential.

Strategies for Achieving the Goal

The strategies should include:

(a) Preparation of an Action Plan for two-year and five-year periods. The Action Plan will be based on the WRCC mandate, priority water resources management issues at the national level and in the river basin. It will be necessary for the Action Plan to emphasize the need to integrate a more balanced view of sectoral perspectives in the programme of investment and operation of all water-related agencies. WRCC will present the Action Plan to the Government for approval.

(b) Preparation and implementation of a Capacity-Building Plan for the WRCC, based on the requirements for accomplishing the Action Plan. The Capacity-Building Plan will include:

  (i) The expansion of WRCC staff from its present size of 12 to 25 persons;

  (ii) Staff selection and management;

  (iii) An internal structure (tied to the mandate and long-term work plan of the Secretariat), including position descriptions and assignments;

  (iv) A WRCC administrative system, including financial, accounting, asset management and record keeping, planning and budgeting, and priority and goal setting; and

  (v) Awareness raising for WRCC members and training for WRCC staff, to be implemented under the project's training component. It is expected that external assistance will be provided in implementing all aspects of the Capacity-Building Plan.

(c) Conducting a training needs assessment for integrated water and resources management training in WRCC and the provinces in the river basin. Preparation of a coordinated training plan to address the priority training needs, drawing on the training modalities identified in the project and the available training budget. The training needs assessment and plan will be submitted to WRCC for approval. Support will be solicited for WRCC to facilitate and coordinate the delivery of training activities under the approved plan as well as evaluate training results, and recommend ongoing training needs and activities.

Performance Measures

The performance measures should include a WRCC Action Plan, a Capacity-Building Plan and the total number of staff trained with advanced technology.

IMPLEMENTATION, MONITORING AND EVALUATION

Human Resources Development

Having established a strategic plan management process, it must be implemented. To do this, alignment of the organization as well as development of the right skills and competencies to match the strategy must be undertaken. This includes:

(a) Building the organization around the critical strategic elements;

(b) Mobilizing and allocating resources to match the strategic plan;

(c) Establishing administrative systems, policies and procedures in accordance with the strategic plan;

(d) Developing a personnel reward system linked to achievement and the strategy; and

(e) Recognizing that the most important task of leadership is to establish and believe in strategic plan management as well as ensure that everyone in WRCC complies.

b. Funding for the Future

Because of the almost complete reliance of WRCC on external funding for continued existence, it is crucial that a clear fundraising and resources mobilization strategy be developed and implemented urgently. Only if fundraising is treated as a fundamental part of operations, and approached in a professional and structured fashion, will there be any chance of WRCC becoming financially secure in the future. If it is to become the leading organization in promoting and coordinating water and water resources development for the benefit of the river basin population, WRCC has a duty to treat fundraising as an essential component of its overall operation. Attaining financial security means that funding must be sought and obtained creatively from diverse sources, including WRCC members, development partners, the private sector, members of the public and others. Because of their respective financial situations, it is not expected that total funding by WRCC members will be achievable in the foreseeable future, although that is the ultimate aim.

For the present, a sensible, rational mix of funding needs to be the target. Nevertheless, WRCC will explore what might be considered as a reasonable goal for funding and in-kind services over the term of the strategic plan management, what the funding and services should be used for, and how long it could take to reach this level of contribution. WRCC accepts the notion that its members should at least own and support the core of its organization. This means the basic structure as defined in the Water and Water Resources Law. The amount of funding needed to cover these commitments must now be determined and decisions made as to whether each WRCC member is able to meet this level of contribution in the foreseeable future. WRCC may need assistance in moving toward the desired funding proportion over the period of this plan.

Mechanisms for Reporting, Monitoring and Evaluation

a. Coordination Framework

Ministries and agencies responsible for the management, exploitation, development and use of water and water resources are:

(a) The Ministry of Agriculture and Forestry. Responsible for the (i) management, exploitation, development and use of water and water resources in agriculture, (ii) prevention and control of flooding in agricultural areas, (iii) surveying and collecting meteorological and hydrological data on the Mekong River, its tributaries and other rivers outside the Mekong River basin, and (iv) preparation, updating and dissemination of the inventory of water sources and river basins;

(b) The Ministry of Communications, Transportation, Post and Construction. Responsible for the management, exploitation, development and use of water and water resources in the fields of communications, transportation, town water supply, urban drainage, river bank protection, and flood prevention and control. In addition, the Ministry is responsible for the collection of hydrological data and hydrographical surveys for navigation purposes;

(c) The Ministry of Industry and Handicrafts. Responsible for the management, exploitation, development and use of water and water resources in the fields of electricity, industry, mining, drainage from industries, and handicraft activities;

(d) The Ministry of Public Health. Responsible for the management, exploitation, development and use of water for rural domestic consumption and health care;

(e) The Ministry of Trade and Tourism. Responsible for the management, exploitation, development and use of water and water resources in tourism;

(f) The Science, Technology and Environment Agency. Responsible for ensuring coordination between different line agencies in establishing rules and regulations pertaining to the management of the environment, and research, scientific and technological services related to water and water resources;

(g) The Lao National Mekong Committee. Responsible for coordinating with line agencies in carrying out studies, and for implementing policies, strategic plans and programmes for Mekong development projects in the Lao People's Democratic Republic that are included in the Mekong River Commission Development Plan, which is being prepared in coordination with countries in the Mekong River basin region, other countries and donors. The Lao National Mekong Committee is responsible for drafting laws and regulations that may be necessary as a result of the Chiang Rai Agreement of 5 April 1995; and

(h) WRCC. Responsible for coordinating with line agencies in drafting strategies and action plans, programmes and regulations necessary for the planning, management, use and protection of water and water resources. It is also responsible for monitoring, control, promotion and reporting on the implementation of activities related to water and water resources.

The above ministries and agencies shall coordinate with the local authorities in the detailed determination of responsibilities and scope of activities within their sectors.

b. Licensing of Water use and Water Resources Development Activities

The relevant ministries will issue regulations on the licensing of medium-scale water use and water resource development activities within their sectors. Local authorities will be involved in licensing medium-scale water use. The Government must approve medium-scale water use that is important. Individuals and organizations wishing to undertake large-scale water use and water resource development activities must apply to the relevant ministries for approval by the Government. Applications for large-scale water use must be accompanied by feasibility and environmental impact studies. The relevant ministries will prepare guidelines on the preparation of feasibility studies and environmental impact studies in accordance with Article 8 of the Environmental Law.

The ministries, WRCC, the Lao National Mekong Committee and local authorities will monitor and control the development of water resources under Notification and Public Consultation on Water Resources Development. All water resources development projects, particularly the construction of dams, irrigation schemes, and the diversion of rivers and streams, must go through a public consultation process with the local population. When a water resource development project is expected to have an adverse impact on the natural or social environment, the ministries and local authorities will issue a notification and inform the public about the project through newspapers, public notices and other means. The ministries and local authorities, depending on the type of project involved, will determine the duration of such a notification, and they will organize public hearings at the relevant location. The developer or its representative will attend such public hearings and provide a detailed explanation and information on the proposed mitigation measures. All costs associated with the public hearing will be the responsibility of the developer.

c. Monitoring and Evaluation

The relevant ministries, WRCC and local authorities are responsible for monitoring and evaluating the implementation of mitigation measures and will report to the Government on a regular basis. There are three types of monitoring and evaluation: (a) regular evaluation; (b) monitoring with prior notice; and (c) evaluating without prior notice.

CONSULTATION MECHANISMS AND INITIAL FINDINGS

People-centred Approach

National policy promotes a "people-centred" approach to achieving the objective of "Growth with Equity". In line with this approach, the local communities, non-governmental organizations (NGOs) and stakeholders in socio-economic development activities have been encouraged to participate. This approach is relatively new and thus unfamiliar to most national agencies. However, there is no established practice for dealing with this matter. Currently, regulations on environmental impact assessments for medium- and large-scale development projects are enacted; consultation workshops are organized at the community, district, provincial and central levels before decisions are made. In the irrigation and rural water supply subsectors the selection of projects takes into account the readiness and willingness of a community to participate in project implementation, operation and maintenance.

NGOs have contributed to the development of community irrigation schemes and rural water supplies in remote areas. The Department of Irrigation has estimated that there are about 150 to 200 irrigation schemes, ranging from 5 ha to 100 ha, which are supported by NGOs. NGO-sponsored water supply schemes have utilized rainwater, surface water and groundwater supplies.

Consultation Mechanisms in Place and Future Developments

a. Framework for Participation

Based on the Constitution, public participation issues have been provided for within a number of laws, decrees, notices, guidelines etc., regarding sector-based specific requirements. In the Lao People's Democratic Republic, all key legislation related to land (1996), water (1996), mining (1997), electricity (1997) and forestry (1997) takes account of the need to minimize adverse social and environmental impacts. The Mining Law (1997) also addresses the social need to avoid economic hardship for project-affected persons. In addition, in 1999, the Government approved an Environmental Protection Law that made provisions for the need to consult with project-affected persons during the environmental assessment process. The Government also developed a draft resettlement policy in 1997, which states "that the population participates in the consultation, planning and design process of their new settlement and production areas." Some of the specific public participation issues related to each sector are detailed below:

(a) In the environmental sector, the Environmental Protection Law specifies that all persons and organizations residing in the Lao People's Democratic Republic have an obligation to protect the environment. The provision of relevant data and information, raising public awareness and providing education to the citizens on the importance of the environment are encouraged. In the framework of institutional arrangements, the environmental management and monitoring processes are included in all of the organizations, from the local grass-roots level to the top ranking central level - village administrations, environmental management and monitoring units at the district, provincial, municipal/special zone and ministerial levels as well as the Science Technology and Environment Agency as a whole. In addition, Article 6 of Final Draft of Environmental Impact Assessment Guideline of the Lao People's Democratic Republic specifically states the environmental commitments and obligation of different partners, such as project developers, project-affected people and other stakeholders. Principles, approaches and methodologies to implement the Guideline related to the public participation are also provided;

(b) The Water and Water Resources Law states that Water and Water Resources are the property of the people of the Lao People's Democratic Republic as a whole. Water and water resources developers must contribute funds for water sources and water resources. If water sources development required human resettlement, the developers or projects must make adequate arrangements and provide funding for such resettlement. In addition, within the framework of the WRCC, which is playing the role of advisor to the Government on water and water resources related issues, some key documents have been drafted, including:

   (i) A Draft Decree on implementing the Water and Water Resources Law as well as the establishment of responsibilities of different ministries, agencies and local authorities with regard to the management, exploitation, development and use of water and water resources;

   (ii) A Draft Policy on Water and Water Resources with the objectives of ensuring that the management, exploitation, use and development of water and water resources is sustainable, equitable and supports the goals of socio-economic development and environmental protection of the Lao People's Democratic Republic; and

   (iii) A Public Awareness Programme that has, as its main objectives, the creation of public awareness of the importance of water and water resources, the vulnerability of such resources and the sense of right and wrong concerning the development, management, use and preservation of water. At the same time, this will allow public involvement to be effectively implemented for the well being of the people and the sustainable use of the water resources. In the drafting process of those documents, a number of broad discussion meetings with different line agencies were organized at all levels throughout the country. That provided the opportunity for those agencies to contribute their views and raise any concerns regarding the drafted documents. The meetings also made possible the collection of comments, suggestions, data and information, in order to ensure that the documents were consistent with the prevailing situation of the country.

(c) The Articles of the Forestry Law state that "... the Government promotes the individual and organizations involved in forestry conservation, rehabilitation, reforestation and extension processes by providing policies, regulations and measures, in order to ensure the abundance, values and sustainability of the forestry resources"; the Articles also state that "... After the forestry allocation and classification processes, and identification of forest areas and forest lands, the Government will transfer the rights to the local governments, and then the provincial and municipal authorities will transfer the rights to districts and then to the village administrations to be responsible for such resources. The provincial and municipality authorities, and district or village administrations that share their borders, will be involved in those processes".

(d) The Electricity Law states that the environmental impact assessment process in the construction of electric power plants must take into account the impact on the communities living within the project areas.

A policy identifying the different management responsibilities of the central and district authorities has been adopted. The provincial level is to undertake strategic planning. The district level is to be responsible for planning and budget. The village level will implement plans.

b. Level of Public Participation

In the Lao People's Democratic Republic, there are four levels of public participation:

(a) Information gathering. This involves the systematic analysis of existing social, cultural and economic conditions among directly affected groups of stakeholders (such as farmers or indigenous minorities);

(b) Information dissemination. This refers to the provision of information about a project to all interested parties (stakeholders). Stakeholders cannot genuinely be consulted or participate if they are not fully informed about the objectives of a project;

(c) Consultations. This stage is where decision-makers listen to the views of other stakeholders in order to improve project design prior to implementation, or to make necessary changes during implementation. Consultation should involve the Government, affected parties, donor agencies, mass awareness organizations and NGOs (local and/or international); and

(d) Participation. This stage is an extension of the consultations where directly affected groups become joint partners in the design and implementation of projects. They participate in "making the decisions".

The lessons and experiences acquired by each sector related to the creation of awareness and implementation of public participation policy are different. However, in general, these activities are new and have not been given priority, since each sector has been obliged to concentrate on project implementation with the aim of meeting the urgent demands of society.

CONCLUSIONS AND RECOMMENDATIONS

In general, progress in the integration of water resources management into economic and social development plans in the Lao People's Democratic Republic was found to be good through the assessment of various integration elements, such as progress in water resources assessment, monitoring and planning, in water supply and sanitation and in mobilization of community participation. The experiences of the Lao People's Democratic Republic show a wide variation in the modalities and methodologies used for integration. Among them, the most commonly accepted methodologies were long-term development plans and medium-term development plans (10 years). The use of water resources for economic growth, food security and poverty alleviation was found to be the most common top priority in water resources development.

Information received was extensive with regard to the legal and institutional framework, which can be considered an indicator of the important progress made in those areas. SPM results indicated that planning information was the most important aspect in coordination at the line ministries, while water resources information was considered to be most essential at the provincial/ basin levels. SPM also provided some guidelines with respect to the needs of cooperation and technical assistance among the eight priority areas identified in the National Economic Development Plan. The overall SPM in the WRCC experience suggested that the process of integrating water resources management into economic and social development plans be viewed on two levels: the overall level in connection with national economic and social development, and the sectoral level for integrated water resources management. At the overall level, five elements were identified: water planning coordination, policy and registration development, data and information management, training, and public awareness for integrated water resources management. At the sectoral level, the process was seen as comprising the two main elements of development and management, supported by their major programmes, with final integration into a national water resources action programme.

5

Management of Water Resources in Malaysia

INTRODUCTION

Malaysian Government through the Economic Planning Unit (EPU), Department of Irrigation and Drainage (DID), and Malaysian Water Partnership (MyWP) initiatives has conducted a two days workshop on Strategic Planning and Management for the Langat River Basin from 21st to 22nd July 2003.

Altogether, over 70 senior experts from various national agencies, local authorities and NGOs related to the Langat River basin attended the workshop including the Executive Committee of MyWP and Mr. Ti Le-Huu, Economic Affairs Officer of the Water Resources Section of the United Nations Economic and Social Commission for Asia and the Pacific (UNESCAP).

The Langat River basin was chosen as the prime/case study for this project since this basin exhibits interesting conflicts among the stakeholders. Among these are (i) one of the most studied river basin in Malaysia, (ii) shared between Federal Territory, State of Selangor and State of Negeri Sembilan, (iii) serve as the core natural river within premier development activities like Putrajaya, Cyberjaya, Kuala Lumpur International Airport (KLIA), Nilai, Ulu Langat and Puchong industrial estates, commercial and housing estates, and at least 4 higher learning institutions, and (iv) there are sensitive environmental issues/landmarks such as incinerator at Broga, water supply dam and intake points as well as North and South Langat peat swamps.

Background of the Langat River Basin

The Langat River basin is a unique basin since it passes through 3 distinct administrative regions of (i) Federal Territory of Putrajaya and Cyberjaya, (ii) State of Selangor, and (iii) State of Negeri Sembilan. The entire basin is slightly less than 2,400 sq. km. with its uppermost stretch originates from the Titiwangsa main range of the State of Selangor and Negeri Sembilan.

The total length of this river has been recorded at 200 km and ends its journey into the Strait of Malacca. In this process it flows through the nation heart of Putrajaya and Cyberjaya. Putrajaya holds the new Central Government Administrative Complex while the Cyberjaya locates the major international IT industries. The Langat River basin base map. For the benefit of the entire study, this basin has also been divided into 10 strategic zones that reflect on the degree of economic and social interests.

Objective of the Workshop

The workshop was meant to identify the various issues underpinning the water resources sectors within the Langat River basin. Five main issues that have been elaborated and discussed, which include:

- Land use;
- Biodiversity and water quality;
- Domestic, industrial and agricultural water supply;
- Floods; and
- Support and capacity-building.

The second objective is to lay out the plan of actions in such a way that the problems are comprehended by the various stakeholders while at the same time enable control measures to be strategically imposed according to short and long terms approaches.

The third objective is to identify the key factors and forces that will ensure the specified action plans are implemental and sustainable within the context of the Langat River basin especially and in Malaysia generally. The final objective is to synthesize the action plans based on the Integrated River Basin

Management (IRBM) plan for the five main issues. This technique speculates on:

- Determination of lead agency;
- Expected outputs; and
- Priority short-term plans.

Finally, all the objectives were gathered and recommendations arisen were strengthened and polished further into acceptable plan of actions within the context of IRBM plan that could be utilized by the three respective administrative regions within the Langat River basin.

Outcome of the Workshop

Issues Identification

Only three (3) sub issues have been prioritized for each main issue. Further analysis shows that some of the sub issues are inter related with several main issues like 'Lack of Political Will', which connects to the 'Inadequate Support and Capacity Building' as well as 'Pressure on Domestic, Industrial and Agricultural Water Supply'. Other sub issues that fall under these categories are 'Lack of awareness', 'River Water Pollution', and 'Improper Land Conversion/ Expansion on Urbanization'.

The analysis also indicates that sub issue of 'Improper Land Conversion/Expansion on Urbanization' is related to three main issues that are 'Poor Land Use Management', 'Floods' and 'Depletion of Biodiversity and Water Quality Deterioration'. This is a clear indication why integrated approach is the favourable step towards holistic planning and management of the entire river basin i.e. IRBM.

The Plan of Actions

The rest of the action plans stand by their own, which speculate that fragmented planning and management are required to be implemented in order to tackle each respective main issue. These indications are the unlikely approach towards holistic measures. To apprehend the outcome towards holistic strategic planning and management approach, the groups are given the benefit to reevaluate and conclude the plan of actions by focusing on Integrated River Basin Management (IRBM) for the Langat River basin.

Integrated River Basin Management (IRBM)

The groups are required to conduct brain storming sessions and further discuss on three major points that include (i) determine the Lead Agency, (ii) expected outcomes, and (iii) priority short term plans. Finally, all groups were required to summarize their findings in the conclusions and recommendations.

Lead Agencies

The 'Land Use' group specified that a river authority should be established namely the 'Langat River Authority'. The chairman shall be rotated among the Federal (Putrajaya) and respective States (Selangor and Negeri Sembilan) representatives. Similar enactments like in Selangor should be set up in Putrajaya and Negeri Sembilan. The group of 'Biodiversity and Water Quality' suggested that a 'National River Basin Authority' should be set up. These should cater not only the Langat River basin area but also the whole country. The Department of Irrigation and Drainage (DID) should be the interim caretaker as well as the IRBM secretariat.

The third group of 'Domestic, Industrial and Agricultural Water Supply' proposed the 'Ministry of Water Resources' as the leader supported by Federal and State Water Agencies.

The group of 'Floods' specified the lead agency comprises DID and Local Authority. Other agencies involved as members are Department of Environment (DOE), Department of Town and Country Planning (JPBD), Land Office, Department of Mineral and Geosciences (JMG), Department of Agriculture (DOA), State Water Agencies (LUAS, PUAS etc.), Ministry of Health, Forestry Department, Public Work Department (PWD), Malaysian Highway Authority (MHA), NGOs, General Public Representatives.

The final group of 'Support and Capacity Building' listed DID, DOE, State Water Agencies, Sewerage Services Department (SSD), JPBD, Negeri Sembilan Executive Council (Exco), Department of

Local Government, Perbadanan Putrajaya and NGOs. The committer should be chaired by Selangor State Government while DID as the secretariat. The Selangor State Government was chosen to chair since most of the Langat River basin is located within Selangor.

Priority Short-term Plans

Three other plans that are inter-connected are (i) 'Land use/ alienation/conversion control' that is related to 'Floods Group' and 'Land Use Group', (ii) 'Reduce solid waste by 2005' that belongs to 'Floods Group', and 'Support and Capacity Building Group', and (iii) 'Prepare awareness programme modules/syllabus in school etc.' that is connected to 'Support and Capacity Building Group', and 'Domestic, Industrial and Agricultural Water Supply Group'.

Benefits of SPM Approach

Among the benefits obtained from the SPM approach are:

- Convenient and straight forward technique to identify problems and propose solutions;
- Issues identification at the lowest level by various sectoral experts;
- Suggest appropriate short and long terms solutions based on each issue;
- Lay generic and specific action plans for each issue;
- Pave way the cross sectoral solutions;
- Collate and integrate the solutions into strategically and manageable measures; and
- Line up integrated plan of actions through holistic means.

In short, the Malaysian Government has implemented some of the recommendations listed by the Workshop using the SPM approach such as:

- Establishment of Ministry of Natural Resources and Environment in March 2004;
- Gazettement of water catchment areas (Cabinet Directive November 2004);

- Formulation of River Law (in process);
- Squatters free by 2005;
- Love Our Rivers Campaign (ongoing);
- Study on Integrated River Basin Management (IRBM);
- River rehabilitation and restoration programmes;
- Establish National Rivers Blueprint and etc.

Challenges of SPM Approach

Among the challenges discovered by employing the SPM approach are:

- The ability of each expert to accept other view point and reach agreeable solutions
- The ability to voice out their points into simple and comprehendible means
- The willingness to be transparent during discussion
- Needs strong and dependable vehicle to bring up the recommendations

Plans for Sustaining SPM Approach

Practical plans that could adduce the sustainability of SPM approach are:

- Reliable technique utilized during discussion (card, matrix, SWOT etc.)
- Create small but effective action group (champion) to implement action plans
- Powerful and continuous sponsor to fund project
- Follow up by experts on periodical/monthly basis
- Establish networking among the partners

CONCLUDING REMARKS

The workshop was able to bring up important issues and propose solutions generated from five identified main issues through implemental action plans. The workshop was conducted by dividing the participants into ten groups denoting two groups per issue. The initial process was to diverge the issues through brainstorming sessions by using various facilitations techniques. Most of the groups utilized card techniques to extract ideas. For choosing the appropriate solutions, action plans and recommendations, most groups used matrix technique.

The concept of Strategic Planning and Management (SPM) introduced in this workshop has enabled the water resources stakeholders within Langat River basin to convene and discuss on the basic issues, solutions and action plans that are vital for the formulation of Integrated River Basin Management Plan (IRBM) in Malaysia. Some of the inputs and outcomes that have been put through in this workshop last year have been applied and harnessed by the Government recently and further improvement on the overall concept of IRBM is moving progressively. With these positive steps, the utilization of SPM technique in Malaysia is heading towards the right track and could be utilized also for the same purpose on other strategic management issues in the country.

6

Strategic Plan on Integrated Water Resources Management in Myanmar

INTRODUCTION

Since water has a direct or indirect relationship with poverty, governance, environment, climate, power, agriculture, floods, food, education and culture etc., society cannot sustain life without managing water wisely and solving water problems. Therefore the Government of Myanmar has formulated plans for developing, modernizing, industrializing and promoting the quality of life for all; to do this, it has laid down four economic objectives for developing the country. One of four economic objectives is "development of agriculture as the base and all-round development of other sectors of the economy as well". The Government has also set the following guidelines on the development of rural areas where the majority of the nation's population lives:

(a) Ensuring good transport in rural areas;

(b) Ensuring adequate supplies of safe drinking water;

(c) Raising the educational standard of the rural areas;

(d) Improving health standards in the rural areas; and

(e) The development of agriculture and livestock farming.

These objectives clearly require support from the water sector, which comprises various agencies related to the use of water for hydropower, irrigation, water supply, fisheries and wastewater treatment. To achieve the development objectives, coordination among such agencies as well as the allocation of specific responsibilities for managing the water resources are essential factors.

National Water Sector Context

The water basin characteristics in Myanmar are quite variable due to the differences in physiographic features. The principal watercourses flowing separately in Myanmar comprise four major rivers, the Ayeyarwady, Sittoung, Thanlwin, Bago, plus their major tributaries such as the Chindwin, Myittha, Mu, Zawgyi, Panlaung, Samon, Myitnge, Mone, Man, Salin, Yaw and Mindon. All rivers, with the exception of the Thanlwin within Myanmar territory and can be considered nationally owned water assets. Their drainage area is spread widely over the country, amounting to some 876.73 million acre/feet (1,082 km^3) of water volume per annum from a drainage area of about 284,800 mi^2. The monthly distribution of river flow closely follows the pattern of rainfall, i.e., about 80 per cent during the monsoon season (May - October) and 20 per cent in the dry season (November - April). The estimated groundwater potential in Myanmar is around 495 km^3 in eight principal river basins in Myanmar.

There are about 200 gauging stations under irrigation department for water level recording and discharge measurement. Some 70 hydrological stations have been installed along the Ayeyarwady, Chindwin, Myitnge, Sittoung, Thanlwin, Bago and Kalandan rivers since 1965 by Department of Meteorology and Hydrology. The department has about 30 discharge stations and 20 sediment discharge stations on main rivers and big tributaries as well as about 15 water quality stations on rivers of Ayeyarwady delta area for measuring discharge and sediment flows and monitoring salt intrusion. These measurement data are valuable for national planning related to water management and, in August 2003, the Myanmar National Committee for the International Hydrology Programme for cooperation with the United Nations Education, Scientific and Cultural Organization (UNESCO) was established for strengthening these activities.

The Ministry of Forestry is responsible for the rehabilitation and conservation of forests and watersheds and for maintaining the stability of the environment, in order to develop the social and economic conditions of the nation, especially in rural areas.

The National Commission for Environmental Affairs (NCEA) was formed in February 1990 to deal with all environmental matters. In March 2004, the Environmental Conservation Committee was set up with the aim of carrying out environmental conservation activities in the country effectively and systematically.

Scope of the Study

The main objective of strategic plan management is to enhance the application of integrated water resources management in the country. In this connection, the components of integrated water and resources management to be studied when formulating a strategic plan are:

(a) Principles of water resources development and management

(b) Operation and management

(c) Water allocation among competing uses and users

(d) Water productivity at the farm, system and basin levels

(e) Financial resources for water source development and management

(f) Conjunctive use of surface water and groundwater

(g) Interactions between irrigation, human health and environment

(h) Public involvement

(i) Capacity-building and human resources development

Experience and effects should be considered when formulating the development and management strategic plan. It is also important to take into account the issues related to the cultural and adverse effects on the environment.

Water Resources Utilization and Challenges

Myanmar is an agricultural country with an abundance of water resources. The agricultural sector is the most basic economic of the State as well as the main source of livelihood in rural areas, since the rural population represents some 70 per cent of the nation's population. The development programmes for the agricultural, livestock breeding and fisheries sectors are included.

At present, the State has been systematically disseminating advanced techniques and support for the development of the nation's economy. Irrigation facilities have been increased since 1988 to reach 163 in August 2004. Dams are now irrigating more than 1 million ha of farmland. In addition to the dams, river water pumping stations, underground water tapping stations and small dams have been built throughout the nation. A total of 265 river pumping projects are irrigating about 150,000 ha of cultivated land. In addition, 7,478 tube wells have been provided to irrigate 36,000 ha of farmland.

Tributaries originating in the western hill region and southern part of the country constitute some 10 per cent of the catchment areas and surface runoff. The hydropower potential of these tributaries is considerable. According to studies by the United Nations and other sources, the hydropower potential of Myanmar is estimated to be as much as 40,000 MW. By 2002, 35 hydropower stations (including 15 medium-scale projects) had been completed with a total estimated generated power of 390 MW, which is almost 1 per cent of potential generated power in Myanmar. The development of the electrical power sector thus can provide a significant contribution to the socio-economic growth of the country. Therefore, electric power projects are being implemented wherever possible to fulfill the electricity needs of Myanmar.

The Development Affairs Committee under the supervision of the Department of Development Affairs has implemented 15,715 projects to supply drinking water to 10,602 villages, using funds form the Government, the Department of Development Affairs and donors at home and abroad. The projects have been implemented with participation by international and non-governmental organizations.

The water sector faces several problems including unusual rainfall patterns in some years, flooding and drought in some of the main agricultural areas of country, the impact of shifting cultivation, illegal logging in water resources areas as well as management conflicts of interest and a lack of coordination within the agencies. The most important challenges include:

(a) Strengthening the legal framework to ensure effective and harmonious integration of water resources management, development and protection activities into the socio-economic development process of the country;

(b) Enhancement and consolidation of the existing systems;

(c) The operation, maintenance and rehabilitation of facilities safely, reliably and efficiently; and

(d) Prioritizing capacity-building needs in order to enhance organizational capacity and effectiveness of the water resources coordination system.

Overall Socio-economic Development Context of the Country

Agriculture (including crop production, livestock, fisheries and forestry) is the most important sector in the country's economy because:

(a) It is the main source of livelihood for about 70 per cent of the population, who live in the rural areas;

(b) It accounts for about 64 per cent of the labour force;

(c) It contributes about 41 per cent of export earnings; and

(d) It contributes about 42 per cent of the gross domestic product (GDP).

Due to the importance of the agricultural sector, the Government has accorded high priority to its development, and numerous irrigation facilities have been implemented during the present decade for irrigation and water supply to monsoon and summer paddy crops. Total irrigated area increased from 1.1 million ha in 1980-1981 to some 2 million ha in 2000-2001. Apart from the agricultural sector, the hydropower subsector is also the most important in terms of economic development and investment. Dry-season irrigation, especially from river pumping projects, has been successfully increased; as a result, power demand is also increasing annually.

The Government is making every effort to be able to catch up with development in other countries around the world. The power generation sector plays a very important role in those efforts and it will contribute considerably to socio-economic growth. Therefore, electric power generation projects are being implemented wherever possible in order to meet the demand for electricity supplies. The Government is implementing plans for the national grid and additional power supply projects. There are 11 major projects being implemented for the power grid and a further 19 projects are planned. Three regional projects are being implemented, with an additional 10 planned for implementation. Myanmar is one of the richest water resources countries in the Association of South East Asian Nations. In view of its current level of water utilization, Myanmar can therefore sell electricity and drinking water if there is a consensus for Global Water Partnership on a regional basis (GWP-SEA).

Opportunities and Threats in Water Resources Development

Myanmar is rich in water resources. The total utilization of the nation's water resources is only about 5 per cent or 45 million acre-feet (56 km^3). It is clear that the physical potential for further development of water resources in Myanmar is substantial. Myanmar has now reached a major turning point in the use of water resources for all-round development of the country. Providing innovative and integrated solutions for sustainable management of water resources to meet national development needs has become a necessity. At the same time, the increasing water use will need careful management to prevent unwanted effects.

Sedimentation is one of the major adverse effects of storage dams and in the lower courses of rivers. Mining and deforestation along the upper reaches of river basins cause serious watershed erosion problems. Transported sediment is reducing the storage capacity of reservoirs and the bed level in the lower reaches of rivers is rising. Consequently, flooding occurs and navigation faces serious problems. Although some nutrients and some sediment are needed to support the aquatic environment, the Government is emphasizing the implementation of the terrace farming system to reduce shifting cultivation.

The development of industry and increasing population density will cause increasing river pollution and health risks for people living close to the rivers. Careful management of groundwater extraction is also required in order to avoid pollution.

Flooding is a major issue, with damage occurring both in urban and rural areas. On the other hand, flooding brings some benefits as it creates rich floodplains and replenishes soil moisture. Flood plain management could be considered as a way to reduce flood damage, with sensitive development being encouraged away from flood-prone areas. The State has given higher priority to providing adequate support and assistance for further development of the fisheries industry in order to enhance food security while protecting the environment. Changes in land use, declining water quality, the construction of dams and other barriers to migration and high levels of capture are likely to be responsible for declining of fish populations. Therefore, increases in water bodies for aquaculture systems such as reservoirs, river courses, fishponds etc. should be temporarily stopped.

VISION STATEMENT

The vision statement of Myanmar is "sustainability of water resources to ensure sufficient water quantity of acceptable quality to meet the needs of the population in terms of health, food security, economy and environment."

LEGAL AND INSTITUTIONAL FRAMEWORK

Legal and Institutional Context of Water Resources Management

The participants in the roundtable Workshop on the National Water Vision in Myanmar, held in Yangon under the framework of cooperation with UNESCAP and FAO in June 2003, proposed that the national water Vision Statement be adopted. They also recommended the establishment of a national level coordination body responsible for water-related activities and issues and improved coordination between agencies. It was noted that most of the laws, regulations and legislation set out in the early 1900s needed to be reviewed and amended. All existing laws, legislation, rules and regulations should be reviewed with the objective of enacting a unified water resources law that would allow the adoption of a more effective legal framework for coordination and management of water resources. Even though there is no single umbrella law covering all aspects of water resources, the laws of Myanmar deal with the subject in one way or another. Many issues in the law, particularly the roles and responsibilities of various agencies for specific activities such as water allocation need to be developed. There is an urgent need for the formulation of further legislation or decrees for proper management. The Central Law Organization (CLO) and the Attorney-General have the final responsibility for issuing decrees.

Current Situation and Perspectives of the Legal and Institutional Framework

At present, there are a number of government water supply agencies with varying water pricing policies and little coordination. There is no apex body responsible for the overall management of national water resources in cooperation with the public and private sectors. Therefore, water conservation with appropriate management and planning practices is urgently required in view of the rapid socio-economic development of the country as well as for protection against water-related environmental degradation.

The participants in the roundtable workshop recommended the establishment of a high-level coordination body in accordance with the three priority objectives of (a) sustainable development, (b) a clearer water resources policy, and (c) the mobilization and effective utilization of resources.

A proposal for establishing a Myanmar Water Commission (MWC) had been submitted to the Ministry of Agriculture and Irrigation for official approval. According to the proposal, the commission will comprise a senior minister as chairman and the ministers from water-related ministries as members. It should also establish a working (coordination) committee comprising heads of the departments in the ministries concerned as members.

The suggested duties and responsibilities of MWC are:

(a) To lay down and prescribe the Policies and Guidelines concerning with water and water resources and required assessment;

(b) The preparation of water laws;

(c) The introduction of new laws, and the enforcement and amendment of existing laws;

(d) To act as coordinator for water users of both a consumptive and non-consumptive nature and to give necessary instructions to the different water sectors; and

(e) To take the necessary action where relevant.

The suggested duties and responsible of the working committee are:

(a) To deal with international organizations such as the United Nationals Development Programme, the Food and Agriculture Organization of the United Nations, the United Nations High Commissioner for Refugees, the International Commission on Irrigation and Drainage and GWP-SEA as well as regional and national organizations;

(b) Conduct coordination work between water sectors;

(c) The preparation of standardization works for water quality of each sector of water usage;

(d) To prepare and implement a National Water Vision;

(e) To monitor and make the necessary assessments, e.g., environmental impact assessment (EIA), for water usage;

(f) Recommendations for water and benefit sharing and resolving conflicts over transboundary water usage between neighbouring countries; and

(g) Decision-making and establishing the cause and effect of water use and development of water resources.

Brief Analysis of Strengths and Weaknesses of the Current Legal and Institutional Framework

The freshwater resources in Myanmar are mainly used by the agriculture sector, with small quantities being used for domestic, industrial and other purposes. Although Myanmar has abundant water resources and no scarcity of water at present, proper management and a strong policy on sustainable and continuous development of the economy and the conservation of the environment are required for the security of future generations.

Present organizational arrangements at the national and provincial levels generally support the achievement of national policies, but the current institutional problems in the water sector are mainly related to (a) the lack of coordination and collaboration between agencies within the sector and with those of other sectors and (b) inadequate communication and coordination between the national agencies and authorities. Despite the many Acts, laws and regulations related to the water sector, most require modification. Therefore, they should be reviewed with a view to enacting a unified water resources law in order to promote a more effective legal framework for coordination and management of water resources. Some Acts (such as the Burma Groundwater Act of 1930.) are weak as jurisdiction was greatly limited and no attempts have been made to amend them. In fact, some Acts are no longer applicable and suitable to the present and changing situation.

Others weaknesses in the water sector are limited manpower, scarce financial resources, and a lack of appropriate monitoring facilities, proper and systematic record keeping, and regular monitoring and surveillance of water quality. As for water quality control, basic standards of quality for drinking water were recommended in 1990, but have not yet been approved.

As for watershed management, the Forest Department is the main agency for undertaking management activities. Prior to 1935, watershed management was a part of forest management; currently, the Ministry of Forestry emphasizes conservation of watersheds and it has implemented projects in some watershed areas. It is also planned to extend the projects to other watershed areas. The new Forest Law was also drafted to be in line with the principles of the United Nations Framework Convention on Climate Change, since the previous Forest Law that was promulgated in 1902 did not deal with the principles. The current law is very important for water resources management as it protects water catchment ecosystems. NCEA established an Environment Conservation Committee in March 2004 in line with the Myanmar Agenda 21 requirement concerning more efficient freshwater resources management.

If MWC is established it could act as a national apex body and manage water allocation. The tasks of MWC would not only be coordination but also reviews of existing laws and the institutional framework, the various water users and the environment.

However, MWC will need sufficient funds and a commitment from the top level of the Government. Joint efforts of the Government and local communities will be the key to the success of water resources management programmes. Institutional strengthening, capacity-building and public awareness are essential elements of development work.

Mission and Vision Statements

The mission statement for the water sector is "to establish a beneficial framework and effective mechanism for managing, developing and protecting water and related resources in an environmentally and economically sound manner in order to meet the needs of the people of Myanmar."

This statement, if adopted by the Government, will provide a guiding light towards establishing national strategies for both short- and long-term efforts by all agencies, people and stakeholders towards the common goals of national socio-economic development and environmental conservation.

STRATEGIC PLAN GOALS

Overall Goal

The overall goal is the alleviation of poverty and upgrading of living standards by means of sustainable development of the water and water resources and conservation of the environment. The overall goal incorporates the three following main mission goals:

1. Mission Goal 1 - Manage, develop and protect water and related resources to meet the needs of current and future generations.
2. Mission Goal 2 - Operate, maintain and rehabilitate facilities safely, reliably and efficiently to protect the public investment.

3. Mission Goal 3 - Enhance the organizational effectiveness of the water resources coordination system, and promote capacity-building.

Mission Goal 1 - Manage, Develop and Protect Water and Related Resources to Meet the Needs of Current and Future Generations

To achieve this mission goal, cooperative efforts are necessary in undertaking the work of managing, developing and protecting water and related resources in every sector related to water.

The framework for such coordination should include:

(a) Facilitation and formulation of related laws and regulations

(b) The improvement of the water resources planning system for all-round development including the extension of irrigated land

(c) The improvement of efficiency in water use

(d) The improvement of technology used in water resources management

a. Facilitation and Formulation of Related Laws and Regulations

Current water resources policies and legislation of all water-related agencies concerned need to be reviewed and modified. Further policies and legislation should be developed for efficient water resources management and environmental protection. CLO is the main responsible agency for compiling and drafting new laws and presenting them to the Government for approval.

Strategies for Achieving the Goal

To assist in this process, involvement of all related ministerial agencies and the key stakeholders will be required in order to establish an appropriate mechanism, such as a Working Group for Policy and Legislation. The working group will be requested to prepare a brief plan for effective management, development and protection of water and water resources and, if possible, to present strategies to deal with the critical issues. For each significant issue, the plan should briefly address the following:

(a) The history of the issues;

(b) Current status of the issues;

(c) Recommended strategies, actions and options;

(d) Implication of the proposed actions and intended result; and

(e) External factors such as the social, policy, legal and other aspects affecting achievement of the strategy and progress in addressing each issue.

Based on inputs by key ministries, agencies and stakeholders, CLO in cooperation with the Policy and Legislation Working Group will analyse the issues and establish a framework for the development of new policies, legislation, regulations, guidelines etc., for consideration and approval by the Government. MWC should also become responsible for analysing the issues and their submission, if Government approves the establishment of MWC.

External Factors Affecting Goal Achievement

The most important factors affecting the progress of work are (a) the willingness by all key agencies and stakeholders to participate, (b) the lack of expertise in the field, (c) public awareness and (d) availability of funding.

Cross-cutting Relationship

Regular consultations and the establishment of common targets with all related agencies will be necessary for evaluation.

Performance Evaluation and Monitoring: It will be necessary to prioritize socio-economic development activities in priority areas of the country aimed at meeting poverty eradication goals. Specific target dates with appropriate indicators should be identified for the priority areas, particularly those areas facing water issues.

Performance Measures

The performance measures include:

(a) A comprehensive review report on priority issues in water resources policy and regulation, to be completed within three years;

(b) The formulation of a strategy to develop policies, legislation, regulations, guidelines and other legal instruments to be submitted to the Government for consideration and approval by 2010; and

(c) The establishment of a Working Group on Policy and Legislation Improvement that, after being approved, will become active by 2010.

b. Improvement of the Water Resources Planning System

To coordinate the development of water resources planning, a detailed procedure plan approval and implementation procedure will be required. The plan should be based on the interests of the public and local requirements. Local authorities, the agencies concerned and local communities should conduct detailed project planning in a coordinated manner. The plan will also include the responsibility for monitoring implementation of the plan.

Strategies for Achieving the Goal

Available resources (including manpower, water resources and technology), allotment of funds, and local needs and interests for conducting plans must be considered. It will also be necessary to devise a number of strategies to respond to the priority needs of socio-economic development, particularly stable and sustainable economic growth as well as poverty eradication goals. The following four strategies are mainly adopted for planning purposes:

(a) A pilot river basin study. The Sittoung River basin has been selected for this purpose;

(b) An implementation plan for increasing irrigated farmland for food security;

(c) A plan for promoting the energy sector by implementing hydropower projects; and

(d) A plan for safe drinking water in dry zones and other areas.

External Factors Affecting Goal Achievement

The most important factors affecting achievement of this goal are: (a) willingness of all key stakeholders to participate, (b) public awareness and participation, (c) commitment by the Government and key funding agencies, (d) know-how of local staff,

(e) coordination between local authorities and the agencies concerned, and (f) sufficient staff for implementation.

Cross-cutting Relationships

Regular consultation and establishment of common targets with all related agencies will be necessary for evaluation.

Performance Evaluation and Monitoring

Specific target dates with appropriate indicators should be identified for priority projects.

Performance Measures

The performance measures include:

(a) A report on the development of the Sittoung River basin water resources planning system including detailed procedures and roles of river basin committee and other key participants in the preparation of plans, including plan approval and implementation procedures, within five years;

(b) Increase in the area under irrigation. The ratio of irrigated area to net sown area shows a rising trend, having increased from 18.2 per cent in 2000-2001 to 18.8 per cent in 2001-2002. The long-term target plan for the increment of the irrigated area is up to 25 per cent by 2015;

(c) Promotion of the energy sector including (a) electricity demand and supply management, including types and sources of energy and expansion of transmission system, and (b) the development of hydropower potential. At present, 29 projects are under consideration, of which 19 projects are for the national grid and 10 are regional; and

(d) The implementation of 15,715 safe drinking water projects in 10,602 villages under a 10-year plan.

c. Improvement of Water-use Efficiency

Coordination among authorities, agencies and local people to support the planning and development of water resources projects efficiently in order to meet socio-economic development requirements.

Strategies to Achieve the Goal

The strategies for achieving the goal include:

(a) Improving river basin watershed management

(b) Implementation of rehabilitation projects in existing irrigable areas for saving irrigation water

(c) Improving public participation and awareness

Watershed management projects on sustainable use of water resources have been carried out by the Forest Department in some major watershed areas. The projects included reforestation, establishment of community forests, and training for staff and local residents. To save irrigation water, it is essential to carry out rehabilitation work on the existing irrigation system instead of implementing new projects. This includes the adjustment of cropping patterns, water supply scheduling, water pricing, and the initiation of sprinkler and drip irrigation systems. Programmes that promote irrigation efficiency directly affect the farmers through the reduction of energy use, labour and costs as well increases in production. Public participation and the establishment of Water User's Associations (WUAs) will directly increase water-use efficiency. Conducting training and extension works could improve awareness and expertise among local communities with regard to saving water.

External Factors Affecting Goal Achievement

The most important external factors include (a) willingness of the public to participate in demand management, (b) the lack of sufficient expertise in advanced technology and facilities, (c) limited numbers of staff in the agencies concerned, and (d) insufficient allotment of budgets and other necessary funds.

Cross-cutting Relationships

Regular consultation and establishment of common targets will be necessary for evaluation.

Performance Evaluation and Monitoring

Specific projects should be selected according to priority so that their results can be directly linked to service improvement according to requirements.

Performance Measures

The Forest Department has already developed a 30-year Master Plan (2001/02 to 2030/31), aimed at undertaking forest conservation and restoration. By the end of the 30-year plan, the following work will have been completed:

(a) The formulation of land-use plans for all states and divisions of the country with three revisions;

(b) The establishment of 199,355 ha of watershed plantations;

(c) The completion of 450 gully control units, 182.88 km of terrace farming, 274.32 km of percolation and diversion ditches, and 914.4 km of contour bunds; and

(d) The establishment of 295,431 ha of community forest.

The Irrigation Department has to implement rehabilitation work for the North Nawin Dam Project, the Zawgyi Irrigation Scheme and the South Nawin Dam Project within five years. The department also has to conduct training courses on irrigation water management for staff and local farmer as well as prepare reports on training programmes for each year and proposals for the formation of WUAs in appropriate areas.

d. Improvement of Technology used in Water Resources Management

The application of science and technology to water resources management will facilitate decision-making, improve coordination and enhance water-use efficiency.

Strategies to Achieve the Goals

The strategies include:

(a) Coordinating the development of detailed water resources data and management, including an assessment of current practices and issues as well as recommendations for more applicable technology;

(b) A review update, dissemination of existing data, and the preparation and maintenance of a database system and information system for project planning purposes;

(c) A review of the need for additional meteorological and hydrological data within the available budget to enhance the quantity and quality of data for calibration and future application; and

(d) Making recommendations for site locations, selection of equipment and supervision of the installation of new equipment.

External Factors Affecting Goal Achievement

The most important external factors affecting the projects include (a) willingness of all key agencies to participate in the establishment of coordinated data systems, (b) the lack of a data communication system, (c) the lack of sufficient expertise in the use of advanced technology, and (d) insufficient budget for investment in advanced equipment.

Cross-cutting Relationships

Regular consultations and the establishment of common targets will be necessary for information management improvement.

Performance Evaluation and Monitoring

It will be necessary to prioritize activities with target areas and dates so that their results can be applied by decision-makers in the implementation of policies and strategies in water resources management.

Performance Measures

The performance measures include:

(a) A comprehensive review of priority issues in water resources policy and regulations as well as existing practices in data information systems to be completed by the relevant agencies; and

(b) The introduction of a strategy for implementing database and management systems for water and water resources projects.

Mission Goal 2 - Operate, Maintain and Rehabilitate Facilities Safely, Reliably and Efficiently to Protect the Public Investment

Mission Goal 2 focuses on operation, maintenance and rehabilitation, to ensure that existing facilities continue to provide project benefits. These facilities provide power and water supply delivery systems that serve agricultural, municipal in addition to power generation, recreation, fish and wildlife benefits, and flood control. Ensuring that facilities are safe, cost effective and reliable means that they can be operated efficiently and effectively in providing project benefits while protecting public health and sustaining the environment.

OPERATION, MAINTENANCE AND REHABILITATION

Strategies for Achieving the Goal

While operating facilities to provide water, power, recreational use, flood control, and fish and wildlife benefits, it is necessary to maintain system reliability and promote water-use efficiency to ensure continuous and effective operating systems. To accomplish this goal, the following strategies are important:

(a) Establish a working group for each relevant water and water resources agency to review practices, and ensure that current operations and maintenance practices are efficient and effective; and

(b) Address the problem of ageing infrastructure, carry out research on infrastructure materials (e.g., geotextiles and soil hardening agents such as Mg-white for seepage control) in order to detect potential maintenance and safety constraints. Future research will focus on developing cost-effective repairs and maintenance procedures.

External Factors Affecting Goal Achievement

Natural events such as flooding, drought, earthquakes and fire can have an adverse impact on facilities. Funding limitations may also restrict accomplishments. Changes in regulations and policy can have an impact on the ability to provide project benefits such as changing cropping patterns and decision-making on priorities for power components and irrigation components in multipurpose projects.

Cross-cutting Relationships

In accomplishing this goal, it will be necessary to coordinate effective operations as needed within related agencies.

Programme Evaluation and Performance Improvements

The establishment of indicators and the monitoring of incidents will be necessary for effective evaluations of performance improvements.

Performance Measures

The performance measures include:

(a) Reporting by the working group of each water-related agency on emerging issues in the operation and maintenance of existing facilities in the water sector by 2006; and

(b) Compiling reports from each water-related agency and proposing a strategy by 2007 to improve efficiency and effectiveness of existing facilities by a special committee, comprising members from water-related agencies.

Mission Goal 3 - Enhance Organizational Effectiveness of the Water Resources Coordination System and Promote Capacity-building

Mission Goal 3 focuses on enhancing the organizational effectiveness of the water resources coordination system and on promoting capacity-building. To achieve this goal, a productive and effective organization is necessary. If MWC is established, it will become the apex body for controlling, managing and improving coordination and support of work for water and water resources in Myanmar. Before establishing MWC, all the water related agencies need to improve their services for the people and particularly top decision-makers in the water sector and the government. In March 2004, Ministry of Foreign Affairs formed an "Environmental Conservation Committee"; 10 task forces were established under this committee to consider river basin management works as well as other water-related issues and environmental conservation works.

(1) Strategies for Achieving the Goal

There are two main strategies for achieving this mission goal:

- Improvement of coordination among relevant agencies. This includes:

(a) reviewing the existing organization capacity, duties and responsibilities of water-related agencies,

(b) establishing an efficient communications system for improving coordination and collaboration, and

(c) taking follow-up action on approval for establishing the Myanmar Water Resource Commission.

- Preparation and implementation of a capacity-building plan, which will include:

  (a) selection of staff for specific fields,

  (b) training,

  (c) an administrative system including financial, record keeping planning and budgeting, and

  (d) raising awareness of staff. External assistance can be expected in implementing training for transferring knowledge and advanced technology. In addition, university curricula for water resources management and engineering programmes should be strengthened.

External Factors Affecting Goal Achievement

The most important factors affecting the progress of achievement are (a) a lack of funds, (b) limited external assistance in terms of expertise and technology transfer, (c) a shortage of staff for sharing training and administration work, and (d) a lack of facilities for a communications network.

Cross-cutting Relationships

A coordinated effort is required within the water-related agencies and with universities to strengthen the curricula for training and programmes.

Programme Evaluation and Performance Improvements

A programme evaluation of current and future capacity performance in existing organizations is required. This evaluation will make it possible to build upon current and future needs for skills and disciplines.

Performance Measures

The performance measures include a report on the review of existing organizational capacity, the preparation of a proposal in 2005 for new organizational capacity and the preparation of a capacity-building plan, including training programmes, after approval is given for MWC to be established.

IMPLEMENTATION, MONITORING AND EVALUATION

Identification of indicators and benchmarks for performance measurements and expected time frame Myanmar, in cooperation with FAO and ESCAP, launched a programme in 2003 to develop the "Myanmar Water Vision". The Myanmar Water Vision has since recommended that a high-level "Myanmar Water Resources Commission" be established. This recommendation is now in the process of being approved by the Ministry of Agriculture and Irrigation.

In March 2004, important progress in the improvement of water resources management was achieved, especially following the establishment of the "Environmental Conservation Committee". Under the guidance of this committee, 10 special task forces were set up, based on the country's main river systems, coastal areas and forest conservation regions.

The most important achievements were related to better coordination of the water sector agencies and departments in water resources management, with increased interest and attention of the Government. To make the National Water Vision a reality, it will be necessary for implementation to be carried out in accordance with established strategic plan management. Myanmar has to face and overcome the challenges of adequate investment, human resources development and financial support from international organizations. In this case, investment should be made after priority is given to a study of resources development for accomplishing integrated water and resources management. For this purpose, it is recommended that the Water Resources Commission play the key and principal role in the realization of a strategic plan on integrated water and resources management. Its fundamental activities would be (a) proper resource mobilization, (b) enhancing participation by all stakeholders and (c) the organization of funding for water resources development.

In order to monitor the implementation and coordination of activities, the following priority indicators need to be taken into consideration:

(a) Most of the storage dams are located on first and second order tributaries of main rivers, causing no hindrance to the flow in the main channels and dually minimizing environmental impact. The Ministry of Agriculture and Irrigation, in cooperation and in conjunction with the Ministry of Forestry, has been actively engaged in reforestation of watersheds with a view to preventing sedimentation in the reaches. It is recognized that the problems of forest area reduction and river sedimentation should be one of the priority indicators for monitoring to be able to maintain the current situation;

(b) So far, no problems have surfaced with respect to shared rivers with neighbouring countries. However, with the anticipated integrated water resources management in the future, it might be judicious to handle the accompanying legal environmental and economic aspects with verity and wisdom on a basis of bilateral or multilateral understanding. It was recommended that conflicts between nations be monitored to reduce the chance of conflict or violations of international laws;

(c) As part of sustainable water resources development, monitoring water quality and saline intrusion in delta and coastal areas is required. These are also the priority indicators for monitoring to ensure the condition and quality of the watershed; and

(d) Concerning resource mobilization, the capital outlay in the water resources sector is quite satisfactory for the time being and it is recommended that annual financial resources be increased to ensure effective support for the sustainable development. It is also recommended that resource mobilization should be doubled in five years.

Mechanisms for Reporting, Monitoring and Evaluation

To ensure appropriate and efficient monitoring and reporting, it is advisable to use existing mechanisms in the following ways:

(a) The Department of Meteorology and Hydrology will be responsible for (i) monitoring and reporting water quality, (ii) developing indicators for watershed quality, (iii) surveying and collecting meteorological and hydrological data of the main river systems, and (iv) disseminating and sharing of water resources data with other water sector agencies;

(b) The Irrigation Department will be responsible for managing the exploitation, development and use of water and water resources in agriculture, reclamation and drainage. It will also be responsible for preventing and controlling flooding in protected areas, the collection of hydro-meteorological and water quality data on a project-oriented basis, and disseminating and sharing data with other water sector agencies;

(c) The Department of Hydroelectric Power will be responsible for the management, exploitation, development and use of water and water resources. It will also be responsible for collecting and disseminating hydrological data and hydrographical survey data together with generated firm power for past, present and future projects;

(d) The Department of Tourism will be responsible for the management, exploitation, development and use of water and water resources in the tourism sector;

(e) The Ministry of Industry 1 and 2 will be responsible for the management, exploitation, development and use of water and water resources in industry and mining as well as water quality from industrial waste;

(f) The Ministry of Health will be responsible for the management, exploitation, development and use of water for health care purposes;

(g) The Ministry of Progress of Border Areas, National Races and Development Affairs will be responsible for the rural water supply and domestic consumption. It will also be responsible for water supply in rural and urban areas, water quality and urban drainage;

(h) The Water Resources Utilization Department will be responsible for implementation, development and use of water pumping from rivers and groundwater to be used to irrigate farmland;

(i) The Directorate of Water Resources and Improvement of River Systems will be responsible for the implementation, management, development and use of water and water resources in transportation, the protection of river banks and river systems as well as the collection of hydrological data and hydrographical surveys for navigation;

(j) The Department of Forestry will be responsible for conservation management, development and use of forests in a sustainable manner. It will also be responsible for the reforestation programme and for monitoring deforestation and forest cover reduction in critical areas;

(k) NCEA will be responsible for cooperation between various line agencies in establishing rules and regulations pertaining to environmental management and for providing advice related to water and water resources. It will also be responsible for giving advice according to Myanmar Agenda 21 and for initiating EIA activities that may be necessary during the course of a project's life;

(l) The Environmental Conservation Committee will be responsible for coordinating with the ministries and for supporting and presenting proposals, with recommendations, put forward by the task forces to the Government;

(m) The Water Resources Commission will be responsible for coordinating the various line agencies in carrying out studies and formulating policies, drafting laws, strategies and action plans for the planning, management, use and protection of water and water resources. It will also be responsible for reporting the results of implementing activities related to water and water resources, including watershed quality indicators, to the Government;

(n) The National Planning Department will be a mechanism for monitoring financial resources allocation and utilization for the water sector. It will also be responsible for compiling resources allocation to the water sector and for providing advice or strategies to increase resource mobilization and utilization; and

(o) The Ministry of Foreign Affairs will be responsible for setting up and monitoring a programme for reducing the number of conflicts with neighbouring countries that are against international laws concerning utilization of water and water resources from transboundary rivers. It will also be responsible for keeping track of, and compiling information on, conflicts and violations and for preparing a report together with recommendations for improving/resolving such conflicts.

CONSULTATION MECHANISMS AND INITIAL FINDINGS

The enhancement of public awareness and participation is crucial to the successful implementation of integrated water and resources management in Myanmar. Stakeholders, national entrepreneurs and non-governmental organizations need to make concerted efforts to participate in the development of the agricultural, livestock, power and forestry sectors. To promote organizational effectiveness, especially for local communities, development measures are being taken to ensure successful implementation of the five rural development tasks. Myanmar has an abundant hydropower potential of about 40,000 MW, which is considerably more than the domestic requirement. Until now, only about 1 per cent of this potential has been exploited and developed. Therefore, the Government has decided to invest extensively in the renewable energy sector, particularly large hydropower schemes, to cover urgent domestic needs. The Ministry of Electric Power is enhancing the organization and functions of the Department of Hydroelectric Power in order to extend construction of the countrywide power grid as a long-term plan.

In the meantime, Government is encouraging the development of small or medium-sized hydropower schemes to be installed in completed storage dams under the Irrigation Department, to supply areas far from the current electricity grid as part of the rural development programme. As a small-scale community development, localized hydropower schemes are being formulated at the border areas for domestic consumption of electricity. This is being done with assistance from non-governmental organizations and participation by the private sector, in order to promote the hydropower sector.

The Irrigation Department is carrying out a community development programme to promote private participation in the water resources sector. The programme includes community irrigation schemes such as the construction of small-scale village irrigation systems and village embankment works. These works are formulated under a capital budget allotment together with contributions from the local villages.

NCEA has already launched a programme on EIAs for medium and large-scale development projects with foreign technical assistance. Consultation workshops and seminars have been held with participation by various agencies and departments. Initiation of EIAs is essential to strengthening institutional and human resources for the protection of watersheds. Although regulations or a decree on EIAs have yet to be enacted, NCEA has disseminated the fundamental regulations for public participation in all stages of a project including construction, operation and maintenance.

Consultation Mechanisms in Place and Future Developments

In Myanmar, several agencies, under their respective ministries, remain responsible for the supply and management of water for agriculture, industrial, domestic and sanitation purposes. In reviewing current conditions, many inconsistencies can be identified in the existing mechanisms for coordination of development among the various agencies involved. Different agencies have different Acts, proclamations and laws, but most of them need to be strengthened in order to overcome problems caused by the lack of regulations on water and water resources management and utilization. The supporting role for a high-level Water Resources Commission is needed in the following priority functions: (a) the establishment of regulations for water resources management, (b) a study of a pilot river basin management project as laid down by the Myanmar Water Vision and (c) the harmonization of the relationship between water-related agencies.

The legal enactments concerning water resources development in Myanmar are found either in special laws related to one or more uses of water or in other Acts that contain a section that is related to specific aspects of water use.

Water and Water-related Laws under the Irrigation Department

According to existing legislation, the ultimate ownership of water resources is vested in the Union of Myanmar. The first Act reserving such a right of overall control was the Burma Canal Act, 1905, which regulated irrigation, navigation and drainage in Myanmar. This Act permitted water in all rivers and streams flowing in natural channels as well as lakes and other national still water bodies to be used and controlled for public purposes.

It was evident that the ratio of irrigated area to that of the net sown area is experiencing a rising trend. It increased from 18.2 per cent in 2000-2001 to 18.8 per cent in 2001-2002. The long-term target for the increment of irrigated area is up to 25 per cent by 2015. There are strong linkages between irrigation and poverty alleviation. The following considerations are aimed at reducing the magnitude of the impact of any irrigation intervention on poverty:

(a) Irrigation infrastructure improvement
(b) The development of irrigation water management and allocation
(c) The improvement of the quality of irrigation water
(d) The enhancement of irrigation technology
(e) The selection of appropriate cropping patterns

The water laws or other provisions thereof are contained in the following laws and their amendments:

(a) The Rangoon Water Works Act, 1885

(b) The Burma Municipal Act, 1898

(c) The Burma Canal Act, 1905 as amended by the Burma Act, 1914, 1924, 1928 and 1934

(d) The Burma Embankment Act, 1905 as amended by the Burma Act, 1923, and 1931

(e) The City of Rangoon Municipal Act, 1922

(f) The Underground Water Act, 1930

(g) The Burma Water Power Rules, 1932

At present, a separate and relevant national policy on water allocation does not exist in Myanmar. The main objective of water resources allocation is to make the best use of available resources for the well-being of the people. Water use in Myanmar, in decreasing order of priority, include the domestic, irrigation, hydropower, industry, fisheries and aquaculture sectors. Generally, there is no standing order of an Act or Rules prohibiting the exploitation of either surface or subsurface water. Some of the existing Acts or Rules are almost two-thirds of a century old and are no longer applicable or suitable to the present situation.

CONCLUSIONS AND RECOMMENDATIONS

The most important achievement during 2005 has been the establishment of the Environmental Conservation Committee with the Ministry of Foreign Affairs acting as chairperson and 10 task forces operating under it. The duties and responsibilities of each task force include reporting to the chairperson with recommendations on river systems, coastal zones and forest conservation regions.

Another achievement during the past three years has been the effort by the Department of Development Affairs to sink tube wells for community development in rural areas, financed by public donations. The cost of sinking tube wells for drinking and domestic water supply amounts so far to K 691.94 million in 1,839 villages in dry zone areas and other parts of the country.

Each water-related department in the ministries has formulated a 30-year Master Plan for the development and management of water resources.

The Ministry of Agriculture and Irrigation has implemented several activities during this decade with regard to the improvement of irrigation facilities, efficient management and use of water, water pricing, the initiation of sprinkler and drip irrigation wherever feasible, and the introduction of appropriate cropping patterns.

The pressure of rapid population growth and attendant demands for more food will inevitably strain the country's water resources further. Therefore, it will be necessary to establish a high-level Water Commission as well as an effective national Water Policy covering water laws, disaster preparedness, efficient water use, ecosystem conservation, institutional strengthening and sectoral coordination of all relevant aspects.

7

STRATEGIC PLAN OF THE PAKISTAN COUNCIL OF RESEARCH IN WATER RESOURCES

INTRODUCTION

The Islamic Republic of Pakistan, the second largest South Asian nation, has a total population of 146 million and a land mass of 79.61 million hectares (ha), of which 70 million ha is arid and semi-arid (including 11 million ha of deserts). The country has a great variety of landscapes ranging from the high mountain ranges of the Himalayas, Karakoram and Hindu Kush (HKH region) with interspersed valleys, and the vast rich irrigated Indus plain to the impressively rugged rocky plateaus of Pothwar, Punjab and South West Balochistan.

The country has four distinct climate seasons. April, May and June are extremely hot and dry months. July, August and September are hot and humid with intensive heat and scattered rainfall. The cool and dry period starts at the beginning of October and continues through November. December, January and February are the coldest months of the year. Due to the diversity of the climate, a large variety of crops is grown to support the agricultural economy.

Water Resources

Pakistan's water resources include rainfall, surface water and groundwater. Rainfall is low and irregular. Annual rainfall ranges from less than 100 mm in the south to about 1,500 mm in the north on the southern slopes of the Himalayas. It generally decreases from the Himalayas towards the south, with the lowest rainfall occurring in southern Punjab and upper Sindh. About 70 per cent of the annual rainfall occurs from July to September (the monsoon period). This causes the loss of most of the run-off to the sea without any economic benefit to the country.

The contribution of rain to crops in the irrigated areas (16 million ha) of the Indus basin has been estimated to be 9.24 billion m3 or 7.55 million acre/feet. In non-irrigated areas (barani areas - 4 million ha) rainfall is utilized for rain-fed agriculture and meeting the drinking water needs of the population and livestock. The contribution of rain to crops in the barani areas has been estimated at 7.34 billion m3 (6 million acre/feet).

The surface water resources depend on the Indus River and the five major tributaries, the Jhelum, Chenab, Ravi, Beas and Sutlej, on its eastern side. Since the Indus Basin Treaty between India and Pakistan in 1960, the availability of water to Pakistan has become limited to the three western rivers of Indus, Jhelum and Chenab, which provide about 173 billion m3 of water annually.

The Indus River alone provides 65 per cent of the total river flow, while the share of the Jhelum and Chenab is 16 and 19 per cent, respectively. Rivers in Pakistan have individual flow characteristics but all of them generally start to rise in the spring and early summer, with the monsoon rains and snowmelt on the mountains causing a combined peak discharge in July and August.

The river flows are at their minimum during winters, e.g., November to February, when the mean monthly flows are only about one tenth of those in summer. In addition to the three major rivers, there are numerous small rivers and streams that are only seasonal, with flow depending on rainfall and they carry practically no water during the winter months.

Most of the groundwater resources exist in the almost 1,600 km-long Indus Plain, which extends from the Himalayan foothills to the Arabian Sea and covering an area of 21 million ha. The water is stored in extensive and deep alluvial deposits under unconfined conditions. This aquifer is fast becoming the supplemental source of water for irrigation. The aquifer has been built due to direct recharge from natural precipitation, river flow and continued seepage from the conveyance system of canals, distributaries, watercourses and application losses in the irrigated lands during the past 90 years.

Outside the Indus Plain, the groundwater resources are also available in inter-mountain valleys of Balochistan and North West Frontier Province (NWFP). The national groundwater potential is 81.6 billion m^3.

Main Issues and Challenges in the Water Sector

The water sector in Pakistan faces numerous issues, with the most prominent being:

(a) Shortages of water due to inadequate storage facility;

(b) Mismanagement of water distribution, operation and maintenance of the irrigation network;

(c) Low productivity per unit of water;

(d) Non-adoption of efficient conservation technologies;

(e) Low irrigation efficiency;

(f) Water quality deterioration (inefficient disposal and handling);

(g) Over-pumping and groundwater mining;

(h) Mixing of fresh and saline aquifers with saltwater up-coning;

(i) Lack of legislation on groundwater utilization;

(j) Waterlogging and salinity including disposal of saline drainage effluents;

(k) Pollution and degradation of wetlands;

(l) Sedimentation of storage reservoirs;

(m) Lack of research on ice and snow hydrology;

(n) Non-participation of beneficiaries in project identification, development and management;

(o) A lack of coordination among research and development agencies; and

(p) A lack of public awareness and education about water conservation.

Mandate of the Pakistan Council of Research in Water Resources

The mandate of the Pakistan Council of Research in Water Resources (PCRWR) is to undertake research and development activities with the following objectives:

(a) Organize, coordinate and promote research in various fields of water resources and set up national research establishments in these fields;

(b) Provide financial and technical support to universities and other research institutions for carrying out collaborative research schemes;

(c) Promote research activities on various aspects of water resources among young graduates through the establishment of fellowships in various subjects of water resources in collaboration with various engineering and agricultural universities;

(d) Collect, disseminate and arrange the utilization of information and research processes or results, and publish scientific papers, reports and periodicals related to the activities of the Council;

(e) Accept fees, donations, endowments and gifts for furthering the objects of the Council;

(f) Establish a library in Islamabad covering the subjects related to the activities of the Council;

(g) Establish scientific liaison with other related national and international organizations; and

(h) Undertake any further activities to promote generally the aims and objects of the Council.

SUMMARY OF THE GOALS LINKED TO SOCIO-ECONOMIC DEVELOPMENT TARGETS

Since the establishment of PCRWR, the council has undertaken 35 research and development projects at a cost of PRs. 638,611,000, including six ongoing projects valued at PRs. 319,188,000. A total of PRs. 259,037,000 has been expended on non-development activities. Research conducted on various aspects of water resources under completed research and development projects has contributed towards the solution of various problems on emerging issues. Some of the contributions include:

(a) Development of technologies to harvest water in desert and mountain areas in order to meet the demand for water during the dry season;

(b) Identification of salt-tolerant plants to stabilize sand dunes in the deserts with saline groundwater;

(c) The introduction of low-cost technologies to divert water from streams in mountain areas to provide increasing regular supplies of irrigation water to small-scale farmers;

(d) Exact quantification of the recharge mechanism through delayed action dams on the Karez in Balochistan;

(e) Determination of the optimum value of the drainage surplus for economical design of drainage systems to control waterlogging and salinity;

(f) Motivation of progressive farmers to share irrigation and drainage costs with the public sector;

(g) Determination of crop water requirements to reduce field application losses in irrigated areas;

(h) Development of irrigation schedules commensurate with the rotational irrigation supply system for wheat, cotton, sugar cane, maize and rice in the Lower Indus Plains;

(i) The fabrication of low-cost soil moisture measuring instruments for farming communities;

(j) Solutions that address urgent drinking water quality problems; and

(k) The introduction of leaky dams to recharge depleting aquifers in Balochistan.

These contributions have ultimately provided very useful socio-economic benefits to the local people and helped to enhance the economy of the country.

PCRWR Research and Development Projects/programmes

Since 1964, PCRWR has been engaged in conducting research on various aspects of water resources through its research and development projects/programmes being carried out all over the country. Keeping in view the rapidly decreasing per capita water availability, extended periods of drought during the recent past, deteriorating drinking water quality, depleting groundwater aquifers and other issues needing immediate attention, PCRWR has submitted a number of projects/programmes to the Ministry of Science and Technology.

The current approved and pipeline projects are described below.

a. Ongoing Projects

The Council initiated six projects throughout the country in 2001:

Mitigation of Drought Disaster in the Cholistan Desert by Management of Water Resources

The Mitigation of Drought Disaster in the Cholistan Desert (MDDC) Project has been initiated in March 2001 at a total cost of PRs. 152,620,000 for a four-year period. The project is aimed at providing safe drinking water throughout the year in the Charleston Desert for humans (110,000 million) and livestock (2 million).

This goal has been achieving by collecting rainwater in 70 newly constructed ponds, each with a storage capacity of 4 million gallons, pumping useable groundwater by installing 20 turbines each with a discharge capacity of 1.0 cusec, and desalinating saline water by a reverse osmosis plant with the capacity of 3,000 gallons per day. A total of 1,686 million gallons water annually will be made available for consumption by humans and livestock after the completion of project.

National Water Quality Monitoring Programme

This programme has been launched in April 2001 with a worth of PRs. 39.66 million for 5 years. Under this programme, surface and groundwater quality of 21 cities of four provinces are being monitored on seasonal basis (before and after monsoon). Apart from these cities, 6 rivers and 13 reservoirs are also being monitored.

Rejuvenation of Depleting Aquifers and Propagation of High-efficiency Systems in Balochistan

This project was launched in April 2001 with a budget of PRs. 38,288,000 for three years in Balochistan Province where rainfall is low (50-400 mm), rivers/streams are non-perennial, evaporation rates are very high (2,000 to more than 5,000 mm per annum), and groundwater is the only dependable source of water for domestic, agriculture and industrial uses. The project was aimed at developing/evaluating new artificial groundwater recharge technologies for depleted aquifers, and propagating high-efficiency irrigation techniques such as trickle, sprinkler and bubbler systems. The recharging technologies would make contributions of more than 20 per cent to groundwater aquifers, thus saving surface water from going to waste through evaporation. The change over from low- to high-efficiency irrigation techniques will also provide water savings of more than water.

Upgrading of PCRWR Laboratories and Libraries

This is a short duration project (two years) with a total cost of PRs. 39.9 million. The key objective of this project is to upgrade capabilities/facilities of existing soil and water laboratories and libraries at PCRWR headquarters and its regional offices while keeping in view the requirements of each laboratory and library. Procurements of state-of-the-art equipment have been to enable the research studies to be carried out accurately and efficiently.

Demarcation of Groundwater Quality Zones in the Indus Plain and its Marginal Areas for Sustainable Development and Management of Groundwater (Phase I)

Initially, this project was commissioned in Punjab Province at a cost of PRs. 38,720,000.

The objectives of the project are to demarcate fresh groundwater quality zones in the Indus Plain and its marginal areas, develop groundwater quality maps for the Indus Plain for future sustainable development of groundwater, and develop a groundwater usability map based on soil-crop-climate and water quality status in different areas for optimum cropping.

On-farm Collaborative Pipe Drainage Project

This project, financed by the Water and Power Development Authority (WAPDA) under the National Drainage Programme, is being implemented in Sindh Province. The objectives of the project are to evaluate the feasibility of on-farm pipe drainage on a watercourse command basis with participation by farmers (sharing of capital costs and bearing full O&M costs), and to develop sustainable low-cost, village-level technology for field investigation, design pipe and filter materials, pipe installations, manholes, sump construction and system maintenance.

b. Approved Projects

PCRWR was to launch the following eight research and development projects in various fields of water resources throughout the country in July 2004:

(1) Sustainable Technologies for Efficient Water Management in Irrigated Areas of the Southern Indus Plain (five years-PRs. 39,670,000)

(2) Result Oriented Short Term Research Studies to Improve Water Resources of Mountainous Areas (five years - PRs. 8,010,000)

(3) Water Quality Monitoring in Rural Areas of Pakistan and Installation of Low-Cost Water Conditioning and Filtration Units (five years - RPs. 8,710,000)

(4) Strengthening of PCRWR by Establishing a Geographical Information System, Hydrological Modelling Centre and Water Resource Data Bank (three years - RPs. 32.2 million)

(5) Strengthening of the PCRWR Regional Office, Lahore for Undertaking Research on Irrigated Agriculture (four years-RPs. 30,739,000)

(6) Impact Assessment of Sewerage and Industrial Effluents on Human Health, Water Resources and Agriculture Productivity in Faisalabad (one year - RPs. 7.80 million)

(7) Enhancement and Management of Groundwater Resources in Balochistan (four years - RPs. 38,690,000)

(8) Recharging Depleting Groundwater Aquifers through Rainwater Harvesting (four years - RPs. 17,560,000)

c. Pipeline Projects

The following five research and development projects have already been submitted to the Ministry of Science and Technology for approval:

(a) Mitigation of Desertification for Poverty Alleviation by Integrated Management of Land and Water Resources in Cholistan (four years - RPs. 34,440,000)

(b) Managing Water Resources in Thar Desert for Desertification Control and Economic Uplift of the Local People (four years - RPs. 39,939,000)

(c) Use of Non-Conventional Water Resources for Drought Mitigation (four years - RPs. 195,630,000)

(d) Participatory National Integrated Water Management Programme (five years - RPs. 174,470,000)

(e) Mass awareness programme for water conservation and development

Opportunities and Threats Facing PCRWR

The research efforts of PCRWR are aimed at maximizing economic productivity of resources through coordinated and collaborated research by:

(a) The improvement of existing resource development and management systems;

(b) The protection of the environment against the detrimental efforts of resource development and management; and

(c) Exploitation of untapped resources.

In order to contribute to maximizing the economic productivity of the resources, PCRWR requires specialization in some selected areas in which a breakthrough can be accomplished.

a. Opportunities

The Council has the following opportunities to conduct and promote countrywide research and development activities:

(a) Collaborative research efforts in the fields of irrigation and drainage, which will make substantial contributions to improving the national economy through:

(i) Developing and popularizing high-efficiency irrigation methods for high delta crops;

(ii) Lowering tile drainage costs to maintain an optimum water table in saline groundwater zones;

(iii) Adopting integrated irrigation and drainage management practices to reduce secondary salinization and up-coning of salt water in central doabs; and

(iv) Improving irrigation scheduling for increasing crop yields per unit of water.

(b) The introduction, evaluation and propagation of artificial groundwater recharge techniques in urban areas where aquifers have been rapidly depleting due to over-exploitation of groundwater. This will not only help in mitigating the adverse effects of groundwater mining but also offset the adverse impact of saltwater intrusion and high energy costs.

(c) Research on the following emerging water related issues to improve water availability of the country:

(i) Water conservation through rainwater harvesting technologies in urban areas, reservoir evaporation reduction technologies in hyper arid areas, etc.;

(ii) Utilization of flood water for groundwater recharging in depleted aquifers; and

(iii) Effectively application of sewerage water and saline drainage effluent for agriculture in conjunction with canal water.

(d) Research studies on the following potential aspects in the northern part of the country, which is not only a haven for tourists but also an area of high precipitation and the location of many watersheds and glaciers:

(i) Sedimentation control

(ii) Watershed management

(iii) Water and soil conservation

(iv) Low flow analysis for hydropower generation

(v) Artificial glacier melts

(vi) Water harvesting technologies

(e) Research on dry and wetland areas with regard to making them productive through suitable land and water management practices. This will be helpful in mobilizing the potential of stressed lands to offset population pressure from irrigated agriculture.

b. Threats

In order to gain the fullest benefit from the above-mentioned opportunities, PCRWR has to deal with the following threats:

(a) Overlapping research activities due to a lack of collaboration between research institutions;

(b) The lack of coordination between implementing, operating and research agencies;

(c) The absence of important research visualized by bureaucracy, industry as well as the public at large;

(d) The lack of appreciation of research work due to long gestation periods; and

(e) The absence of attractive career opportunities for research oriented scholars.

LEGAL AND INSTITUTIONAL FRAMEWORK

Historical Development of PCRWR

In view of the added importance of water, resulting from its uneven distribution in various regions of Pakistan, and to meet the challenges that lie ahead, in 1964 the Government established the Irrigation, Drainage and Flood Control Research Council within the Ministry of Natural Resources.

As an autonomous federal research organization, its function was to undertake and promote research activities in the fields related to water resources. One of its objectives was to provide support, on the research side, to WAPDA and other federal and provincial development agencies in the three major thrust areas, i.e., irrigation, drainage and flood control. The Council started functioning in August 1965.

The Irrigation, Drainage and Flood Control Research Council was brought under the control of the Ministry of Science and

Technology in 1970. In 1984, the Ministry of Science and Technology announced a national policy on science and technology, giving high priority to water resources. Particular emphasis was laid on conducting research and development in various fields of water resources at the federal level. Consequently, the Council was reorganized in August 1985, with a wider mandate and more autonomous character to facilitate efficient functioning, under the present name of the Pakistan Council of Research in Water Resources. The Council has now grown into a premier organization, ensuring the enrichment of the water sector, which functions like an artery in the socio-economic structure of Pakistan.

The Council has remained aware of the problems of limited land and water resources as well as the unabating adverse impacts of water logging and salinity, soil erosion and desertification. Hence, it took timely action in establishing the Drainage and Reclamation Institute of Pakistan (DRIP) at Hyderabad in 1974 and the Pakistan Desertification and Monitoring Unit (PADMU) at Bahawalpur in 1982. DRIP started functioning with Dutch technical assistance to help in controlling water logging and salinity through new drainage technology, especially that of tile drainage and skimming wells. Subsequently, DRIP was moved to the Irrigation, Drainage and Flood Control Research Council to conduct applied research.

The main objectives of PADMU were to monitor research activities for the desert areas of Pakistan, particularly Cholistan, and to study the interaction of wind, rain and overgrazing on the ecosystem of the Cholistan desert.

As PCRWR is a national organization, it has to shoulder the responsibility of solving water-related problems of different provinces. In 1986, it created four Water Resources Research Centres for each of the northern areas of Rawalpindi/Islamabad, Punjab and Balochistan. Moreover, PCRWR developed various projects to solve specific problems of water resources management, safe use of groundwater, and the development of economically and viable irrigation and drainage technologies. It also has discussed ways and means of motivating farmers to shoulder the financial and management responsibilities.

The overall control, direction and supervision of the affairs of the Council is vested in the Board of Governors presided of by the Minister/Adviser in charge of the Ministry of Science and Technology. The Council is headed by a chairperson appointed by the Federal Government. The chairperson is assisted by a Technical Advisory Committee under whose guidance technical policies are designed and implemented. General administrative and financial policies are implemented under the instructions of the Ministry.

Current Situation of PCRWR

PCRWR, with its head office at Islamabad, is conducting research through its eight Regional Offices and Water Resources Research Centres located in all four provinces. These research centres are equipped with experienced research staff, computer facilities, and soil and water testing laboratories. For field research and liaison outside Islamabad, the Council has established eight centres.

a. Drainage Research Centre, Tandojam

In 1997, DRIP was renamed the Drainage Research Centre (DRC). DRC covers 40 acres and includes offices, residential buildings and land for field experiments. It has been conducting research in drainage of agricultural land, salinity control, land reclamation, irrigation practices and water management, and irrigation and drainage modelling.

b. Regional Office, Lahore

The Regional Office, Lahore was established in 1986 for effective linkages and better collaboration with WAPDA, Provincial Irrigation Departments, the International Water Management Institute of Pakistan, International Waterlogging and Salinity Research Institute, and the Directorate of Land Reclamation. In addition to conducting research in various aspects of water management, the office has been providing consultancy/technical services to the public and private sectors in selecting suitable sites for the installation of tubewells with the electrical resistivity survey technique.

c. Regional Office, Bahawalpur

The Regional Office, Bahawalpur was established on the foundation of PADMU. The ultimate objective of this Regional Office, in Addition to liaison with line agencies, is to evolve technology for sand dune stabilization and to save fertile border areas from drifting and moving sand dunes. Lessons learnt from visits to the deserts in mainland China were duly incorporated into the research experimentation at Bahawalpur. This office is conducting research in water harvesting techniques, saline agriculture and saline agro-forestry at its field station in the Cholistan desert. The field station was established in 1986 at Dingarh, about 75 kilometres south of Bahawalpur city by acquiring 500 acres from the Cholistan Development Authority.

d. Lysimeter Station, Lahore

Eighteen lysimeters of 3 m x 3 m x 6 m were constructed in 1970 at the Directorate of Land Reclamation, Lahore to conduct studies on soil-water-plant relationship for salinity and waterlogging control. These lysimeters are equipped with locally fabricated well-tested tensiometers for in-situ measurement of soil moisture tension at required depths. Locally fabricated gypsum blocks are also used for measuring of soil salinity. Water requirement experiments on wheat, maize, sugar cane and sunflowers have been completed, as have experiments on the proportionate contribution of surface irrigation and sub-surface irrigation to total crop water requirement.

e. Lysimeter Station, Tandojam

Twelve drainage lysimeters measuring 3 m x 3 m x 5 m have been installed at the Directorate of Land Reclamation campus to study the soil-water-plant relationship for salinity and waterlogging control in the Lower Indus Plain. Irrigation water requirements for wheat, cotton, and sugar cane have been established and crop coefficient curves of these crops for the Penman equation have been developed.

f. Water Resources Research Centre, Quetta

This centre was established in 1985 to carry out research programmes and to provide the necessary infrastructure to undertake, collaborate and coordinate effective research in the water sector in Balochistan. The centre undertakes research studies to help solve the problems of water scarcity, improvement and recharge of the centuries-old indigenous karez system as well as its interaction with delay action dams and artificial groundwater recharging techniques.

g. Water Resources Research Centre, Islamabad

This centre was established in 1985 with thrust areas of environmental engineering (pollution of water supply and wastewater management), sedimentation and soil erosion in hilly watershed areas. It has considered water quality problems and sources of water pollution to ensure safe and clean drinking water for humans.

h. Water Resources Research Centre, Peshawar

This centre was initially established at Gilgit in 1986 to conduct research studies on developing snowmelt streamflow relationships, character analyses of snow and ice hydrology, and related studies in the northern areas of Pakistan where the Indus River system originates. Little progress was achieved by this centre due to its remote location and frequent ethnic disturbances. Thus, it was moved to Peshawar in June 1993. It is envisaged that the objectives outlined above can be attained at the new location with better efficiency and control.

3. Strengths and Weaknesses of PCRWR

a. Strengths

The Council has 90 professionals in various fields of land and water resources. It has five Ph.D. holders, 42 MEng/MSc holders and 43 BSc (Engineering) holders. For field research, the Council has established four Research Centres at Tandojam, Quetta, Islamabad and Peshawar, two Regional Offices at Lahore and Bahawalpur, and two Lysimeter Stations in Punjab at Lahore and Sindh at Tandojam. Almost all the centres have their own buildings and modest soil and water analysis laboratories. PCRWR has three Research Farms (Tandojam, Dingarh and Lahore), five groundwater investigation equipment sets (terameters), two well loggers, three laser levellers and scrappers, drainage machinery (tile-laying machines, gravel hoppers, tractors etc.) and agricultural machinery (zero-drills, bed and furrow shapers, cultivators and dithers).

b. Weaknesses

The weaknesses of PCRWR include:

(a) A lack of financial resources for research, publication and dissemination;

(b) Inadequate qualified manpower in the following water-related fields:

(i) Hydraulic/river engineering
(ii) Sediment transport
(iii) Dam/geo-technical engineering
(iv) Reservoir operation
(v) Remote sensing
(vi) System engineering
(vii) Snow and ice hydrology

(c) A paucity of equipment and laboratories for performing advanced research in the above mentioned fields; and

(d) Inadequate emoluments, fringe benefits and social status of the research scholars.

4. Vision and Mission Statements of PCRWR

To "make Pakistan self-sufficient in food and fibre, and help achieve the goals of providing better health for citizens through efficient and effective laboratory and field-oriented research in areas of development, use, management and quality of water resources".

STRATEGIC PLAN GOALS

Identification of Priority Areas

In view of the challenges faced by the water sector, the following approaches to research and development in water resources have been identified:

(a) Tapping existing unutilized resources, and the development of new and unexplored water resources;

(b) Managing water resources to achieve the goal of maximum production per unit of water used;

(c) Improving the existing institutional set-up to remove constraints to better management of water resources; and

(d) Improving the research infrastructure by undertaking user-oriented research, avoiding unnecessary duplication of research efforts, and better coordination between research bodies, industry, and universities.

The following 10 areas need to be included in the strategic plan:

(a) Desertification - assessment and mitigation

(b) Groundwater development and management

(c) Agricultural water management

(d) Waterlogging and salinity

(e) Erosion and sedimentation

(f) Hydrology and climate change impacts

(g) Dams, storage reservoirs and water

(h) Water quality - assessment and amelioration

(i) Saline agriculture, disposal and re-use of saline water

(j) Drought and flooding

Research on Managing, Developing and Protecting Water Resources

The goals have been set for undertaking research in the following fields of management, development and protection of water resources:

(a) Development of appropriate rainwater harvesting systems for various potential areas;

(b) Promoting the use of non-conventional water resources;

(c) Development of appropriate and economical artificial recharge techniques in various hydrogeological and climatological conditions;

(d) Developing appropriate technologies/practices that allow the use of drainage effluents for crop production;

(e) Developing and popularizing integrated water management systems for water conservation and utilization on a main and secondary canal command basis with user participation;

(f) Optimization of skimming well technology and its extension to farmers for adoption in areas where shallow freshwater overlies saline water;

(g) Studies to determine the level of pollution, and measures to manage the problems of environmental degradation;

(h) Watershed management in urban, rural and natural catchment areas;

(i) Studies of optimizing cropping patterns under various water availability scenarios in different agriculture zones;

(j) Development of low delta and drought resistant, high-yield crop varieties;

(k) The extension of bio-saline agriculture to the field through coordinated efforts in using salt tolerant varieties, conjunctive use of fresh and saline water, halophytes for fodder production etc.; and

(l) Studies of harnessing snow and ice reservoirs in the northern areas.

Research on Operating, Maintaining and Rehabilitating Facilities

It is planned to undertake research in the following areas of operation, maintenance, and rehabilitation of water resources:

(a) Reactivation and updating of the Indus basin model for operational studies and management of IBIS;

(b) Development of computer models for irrigation/water scheduling at watercourse command level;

(c) Studies of the mechanism of groundwater abstraction, and development of a framework of regulatory authority;

(d) Establishment of a Groundwater Regulation Authority at the provincial level;

(e) Preparation of a National Drought Management Plan;

(f) Studies of drought forecasting mechanisms;

(g) Human resources development in the water sector to deal with the challenges imposed by the current water situation;

(h) Development of a mechanism to strengthen the extension programmes in the provincial setups; and

(i) The introduction of a National Mass Awareness Programme.

Enhancing Organizational Effectiveness

Failure to achieve success in addressing any of the water development/management constraints is common to most government institutions. This stems mainly from (a) widespread pre-occupation with routine duties to meet the procedural requirements of outmoded rules and regulations as opposed to fulfillment of substantive professional responsibilities; and (b) widespread pursuit of self-interest as opposed to the public interest. Furthermore, undue political intervention in the day-to-day affairs of the departments requires arbitrary and often unfair decisions to be made by the functionaries, serving as a major disincentive to staff and leading to the adoption of malpractice on an even wider scale. According to the 10-Year Perspective Development Plan (2001-1011), water shortages will approach 21 per cent in 2011, despite planned additional water sector facilities. The situation demands enhanced management facilities including water management activities, reduced losses in the conveyance system (lining of canals), investment in the reuse of industrial and sewage effluents and the harnessing of non-conventional sources.

The already available research and development infrastructure, climatic conditions and land resources in the country are well suited for maximizing agricultural production and handling the multitude of problems faced by the water sector. What is required is the increased commitment of various administrative, developmental and extension agencies to implementing legislation in protecting water bodies and extending research results to users in the form of better and improved crop varieties, fertilizers, pesticides, cropping patterns, efficient water use, management practices etc. The Government should also ensure the formulation and implementation of better pricing and credit policies for the farming communities, which are aimed at providing incentives not only in irrigated areas but also in rain-fed areas of the country.

Achievement of increased water availability at the farmgate as well as high water-use efficiency to enable high crop production per unit of water used are some other targets for ensuring self-sufficiency in food, or at least for maintaining the current status in the future. Most of the problems are already receiving attention from the Government through the implementation of various programmes with assistance from donor agencies that have been very active in following various developments in the sector. The major programmes being implemented include the National Drainage Programme that provides improved irrigation and drainage facilities in irrigated areas, and the Social Action Programme aimed at improving social services including drinking water supplies and sanitation.

The water sector has experienced meagre investment in the past in relation to other commodities, and most of that investment has been in the development of the water sector. Yet, the management of this huge interlaced sector has been limited. In the public sector, there are certain limitations on making very large investments, especially on the management side for operation and maintenance, regulation etc. At present, there is no significant private sector investment in the water sector. Active community participation is needed in achieving an enhanced private sector role. In addition, some incentives must be introduced for scientists and engineers involved in research and development activities, such as the revival of the research allowance. The recently introduced productivity allowance does not benefit the middle level professionals/establishment related to research. Alternatively, an enhanced pay package, similar to that already being offered to academic people, could be considered.

IMPLEMENTATION, MONITORING AND EVALUATION

Identification of Indicators and Benchmarks for Performance Measures

The growing water demand with respect to population and resultant food and fibre needs require concerted efforts to meet present as well as future water shortages. In this regard, all aspects of water resources development and management have to be taken into consideration.

Expected Framework and Necessary Resources

A three-pronged approach towards the formulation of strategies to meet the growing water needs is proposed:

(a) Tapping of existing un-utilised resources and development of new and unexplored water resources;

(b) Management of water resources to achieve the goal of maximum production per unit of water used; and

(c) Improvement of the institutional set-up and better governance of water resources institutions and infrastructure.

Based on the above approach, the strategies are grouped as short-term (1-3 years), medium term (4-10 years) and long term (11-25 years) extending over a period of 25 years. The necessary resources required for this would involve financial allocations for social sector development activities as well as promotion for research and development of science and technology in the country. In this regard, international donor agencies could also be approached, e.g., the United Nations Development Programme, the United Nations Children's Fund, JICA, IUCN-World Conservation Union, and the Food and Agriculture Organization of the United Nations. The inclusion of local investors could provide support for such activities and offer the additional advantage of indigenous technology development.

Action Plan

Though concerted efforts are needed for improving the national research and development infrastructure in the field of water resources, the recommendations also include development-oriented projects and actions. In this regard, the approach defined in the previous section has been adopted for the formulation of an Action Plan to meet the challenges faced by the nation in the water sector, as summarized below.

a. Short-Term Plans (1-3 years)

The following projects are research and development oriented:

(a) Development of computer models for irrigation/water scheduling at watercourse command level (three years - PRs. 60 million)

(b) Reactivation and updating of the Indus basin model for operational studies and management of IBIS (two years - PRs. 40 million)

(c) Studies of the mechanism of groundwater abstraction and development of a framework of regulatory authority (three years - PRs. 20 million)

(d) Studies of optimizing cropping patterns under various water availability scenarios in different agriculture zones (three years - PRs. 50 million)

(e) Preparation of a National Drought Management Plan (three years - PRs. 50 million)

(f) Studies of drought forecasting mechanisms (three years - PRs. 120 million)

(g) Establishment of a drought forecasting mechanism (two years - PRs. 80 million)

The following projects are research and development oriented:

(a) Developing appropriate rainwater harvesting systems for various potential areas (five years - PRs. 300 million);

(b) Promoting the use of non-conventional water resources (five years - PRs. 450 million);

(c) Developing appropriate technologies/practices allowing the use of drainage effluents for crop production (five years - PRs. 300 million);

(d) Optimization of skimming well technology and its extension to farmers for adoption in areas where shallow freshwater overlies saline water (five years - PRs. 150 million);

(e) Lining of canals in saline aquifer areas with appropriate and effective lining material such as geo-membranes (10 years - PRs. 500 million);

(f) Developing and popularizing an integrated water management system for water conservation and utilization on a main and secondary canal command basis with user participation (five years - PRs. 900 million);

(g) Watershed management in urban, rural and natural catchment areas (10 years - PRs. 20,000 million);

(h) Studies to determine the level of pollution and measures to manage the problems of environmental degradation (five years - PRs. 40 million);

(i) Developing low delta and drought-resistant, high-yield crop varieties (10 years - PRs. 50 million);

(j) Promoting saline fish culture in saline aquifer areas. This would include feasibility studies, research and extension to the field (five years - PRs. 50 million);

(k) Extension of bio-saline agriculture to the field through coordinated efforts in pursuing salt-tolerant varieties, conjunctive use of fresh and saline water, halophytes for fodder production etc. (five years - PRs. 100 million);

(l) Implementation of recharge programmes in stressed groundwater areas/basins through pilot projects (10 years-PRs. 500 million);

(m) Development of appropriate and economical artificial recharge techniques in various hydrogeological and climatological conditions (five years - PRs. 100 million);

(n) Studies and feasibility studies of harnessing snow and ice reservoirs in the northern areas (five years - PRs. 200 million);

(o) Studies on optimization of irrigation systems to minimize sedimentation in canals and watercourses (100 million PRs.-5 years);

(p) Studies of appropriate watershed management techniques suited to local conditions in various physiographic regions (five years - PRs. 100 million);

(q) Strengthening of the existing institutional infrastructure to improve coordination and deal with present and future challenges in the water sector (seven years - PRs. 500 million);

(r) Human resources development in the water sector in order to face the challenges imposed by the current water situation (seven years - PRs. 500 million);

(s) Strengthening of extension programmes in the provincial setups (seven years - PRs. 300 million); and

(t) National Mass Awareness Programme (five years - PRs. 500 million).

c. Long-Term Plans (11-25 years)

The following projects are development oriented:

(a) Continuation of the On-Farm Water Management Project, with more emphasis on lining distributaries, watercourses and minors with appropriate lining material (22 years - PRs. 3,000 million); and

(b) Construction of major and minor multi-purpose dams for storage and power generation (22 years).

Mechanisms for Reporting, Monitoring and Evaluation

PCRWR maintains its monitoring and evaluation unit comprising experts in various water-related areas at the PCRWR headquarters in Islamabad. The Council's research and development activities, which are being implemented all over the country through its six regional centres, are monitored and evaluated regularly for planning/reporting and perusal of the Government's comments. The coordinated research efforts with national and international organizations are also monitored and evaluated by PCRWR headquarters. The reporting period varies depending upon the activity; such reports are published in the form of monthly, quarterly and annual progress reports, among others.

CONSULTATION MECHANISMS AND INITIAL FINDINGS

Consultation Mechanisms in Place and Future Developments

At present, PCRWR is providing consultation facilities in following areas of water resources:

a. Water quality testing, management and technology development PCRWR water quality laboratories in six centres are providing water-testing facilities to the public, the Government and private organizations at nominal charges. In addition, quality management facilities are being extending to various agencies/local development authorities such as the Water and Sanitation Authority. Technology for various organizations such as water testing kits, water treatment units etc. is also being designed.

b. Groundwater Investigation PCRWR provides services to government/private entities for groundwater investigation all over the country. For this purpose, the Council has acquired sophisticated equipment and investigation tools/ software in addition to qualified experts and trained staff.

c. On-Farm drainage facilities. The PCRWR Drainage Research Centre, Tandojam is providing consultancy services to farmers on the installation of drainage systems in Sindh Province. In addition, farmers are being provided with consultancy services in irrigation and land management activities.

d. Consultation on high-efficiency irrigation system. PCRWR has developed a locally pressurized irrigation system that is being promoted among farmers for wide-scale adaptation. PCRWR also provides consultancy services for designing and installing pressurized irrigation such as sprinkler and trickle irrigation systems.

e. Local manufacturing of soil and water instrumentation

PCRWR is manufacturing and supplying various soil and water monitoring instruments/ devices such as gypsum blocks, tensiometers, rain gauges, evaporation pans, potable rain guns etc.

f. Hydrologic investigation. PCRWR also has the capability for provide consultancy in surface hydrology, soil and water conservation, desertification control, flood management etc.

g. Future developments. In future, the Council intends to extend its services to other areas of water resources including hydrologic modelling, water resources management, a geographical information system and water resources database management. To accommodate this plan, a Hydrological Modelling Centre and a Geographical Information Centre are being established at the Water Resources Research Centre, Islamabad. In addition, more field research and demonstration stations are being established for wide-scale dissemination of research findings and the creation of awareness as well as imparting training among water users/farmers.

Initial Findings of Recent Efforts

The Council has made significant achievements through its consultation and technology transfer/dissemination activities. Further research and dissemination programmes have been planned to extend activities in other areas including the northern areas, NWFP, the Pothwar region, the lower Indus basin etc. for water quality management, rainwater harvesting, land and water management, groundwater recharge, drainage, glacier and snow monitoring etc.

CONCLUSIONS AND RECOMMENDATIONS

The conclusions and recommendations are:

(a) Declare PCRWR as a focal point organization at the federal level. A joint forum of water resources research institutes, technical universities and line agencies working in the water sector should be established to watch the national interests;

(b) Keeping in view the importance of research and development activities in the water sector, there is a need to develop capacity-building of PCRWR to handle present and future challenges in order to contribute to the development and management of water resources;

(c) Currently, there is lack of coordination among research institutes, industry and line agencies. Therefore, PCRWR should involve various research organizations, industry, universities and end-users in implementing the Action Plans;

(d) Keeping in view the current and future water shortage in Pakistan, the country has to learn to live with such shortages. Therefore, PCRWR should promote a culture of water conservation through mass media campaigns; and

(e) Currently, PCRWR rarely avails itself of the opportunities to share its experiences/ knowledge with regional and international water-related institutes/agencies. This gap should be closed by developing close linkages with those institutes/agencies in order to meet future challenges in the water sector.

Bibliography

Adelberg, Edward A.: *Papers on Bacterial Genetics*, Boston, Brown and Company, 1960.

Angel, Bradley: *Toxic Threat to Indian Lands: A Greenpeace Report*, San Francisco, Greenpeace, 1991.

Bailey, Conner and Charles E. Faupel: *Environmentalism and Civil Rights in Sumter County*, Colorado, Westview Press, 1992.

Bailey, Ronald: *Liberation Biology: The Scientific and Moral Case for the Biotech Revolution*, Amherst, Prometheus Books, 2005.

Barnum, Susan R.: *Biotechnology: An Introduction*, Belmont, Thomson/Brooks/Cole, 2005.

Burch, William R. Jr.: *Human Ecology: An Environmental Approach*, Belmont, Duxbury Press, 1994.

Campbell, R. R., and Wade, J. L.: *Society and Environment: The Coming Collision*, Boston, Allyn and Bacon, 1972.

Chrispeels, Maarten : *Plants, Genes and Crop Biotechnology*, Sudbury MA, Jones and Barlett Publishers, 2003.

Cox, Earl D. : *Beyond Humanity: Cyber Evolution and Future Minds*, Roackland, Charles River Media, 1996.

Daubenmire, R. F.: *Plants and Environment*, New York, Wiley, 1947.

Dixon, Dougal: *After Man-A Zoology of the Future*, New York, St. Martin's Press, 1981.

Dodds, John H.: *Plant Genetic Engineering*, New York, Cambridge University Press, 1985.

Edwards-Rees, D.: *The Service of Youth Book*, Wallington, Religious Education Press, 1943.

Erickson, Kai: *A New Species of Trouble: Explorations in Disaster, Trauma, and Community*, New York, W.W. Norton, 1994.

Gelobter, Michael: *The Distribution of Air Pollution by Income and Race*, Berkeley, University of California, 1989.

Gitman, Lawrence J.: *Principles of Managerial Finance*, MA: Addison Wesley Longman, 2000.

Hill, David: *The Quality of Life in America; Pollution, Poverty, Power, and Fear*, New York, Holt, Rinehart and Winston, 1973.

Johnson, Hazel: *Financial Institutions and Markets: A Global Perspective*, NY, McGraw Hill, 1993.

King, R.B., Mihony, G., and Sheldon, J.K.: *Practical Environmental Bioremediation*, Lewis Publishers, Boca Raton, 1992.

Kruvant, W.J.: *People, Energy, and Pollution*, Cambridge, Ballinger Publishing, 1975.

Martin, Justin: *The Man Behind Money*, Perseus Book Group, 2000.

Matthews, J. E.: *Working with Youth Groups*, London, University of London Press, 1966.

Milson, F.: *Growing With the Job*, London, National Assocation of Youth Clubs, 1968.

Murray, David: *Seeds of Concern: The Genetic Manipulation of Plants*, Sydney, University of New South Wales, 2003.

Petty, Keown, Scott, and Martin: *Basic Financial Management*, Prentice Hall, 1993.

Primrose, S.B.: *Principles of Gene Manipulation*, London, Blackwell Scientific Publications, 1989.

Retzer, W.J.: *Biotechnology Workbook*, Englewood Cliffs, Prentice Hall, 1991.

Schall and Haley: *Introduction to Financial Management*, McGraw-hill, 1988.

Shetty, Kalidas: *Food Biotechnology*, New York, Dekker/CRC Press, 2005.

Sontheimer, Sally: *Women and the Environment: Crisis and Development in the Third World*, New York, Monthly Review Press, 1991.

Vallette, J.: *The International Trade in Wastes: A Greenpeace Inventory*, Washington, Greenpeace, 1989.

Walden, Richard: *Genetic Transformation in Plants*, England, Open University Press, 1988.

Zupan, J.M.: *The Distribution of Air Quality in the New York Region*, Baltimore, Johns Hopkins University Press, 1973.

Index

□□□